OECD DOCUMENTS

Forestry, Agriculture and the Environment

PUBLISHER'S NOTE

The following texts are published in their original form to permit faster distribution at a lower cost.
The views expressed are those of the authors,
and do not necessarily reflect those of the Organisation or of its Member countries.

ORGANISATION FOR ECONOMIC CO-OPERATION AND DEVELOPMENT

ORGANISATION FOR ECONOMIC CO-OPERATION AND DEVELOPMENT

Pursuant to Article 1 of the Convention signed in Paris on 14th December 1960, and which came into force on 30th September 1961, the Organisation for Economic Co-operation and Development (OECD) shall promote policies designed:

— to achieve the highest sustainable economic growth and employment and a rising standard of living in Member countries, while maintaining financial stability, and thus to contribute to the development of the world economy;
— to contribute to sound economic expansion in Member as well as non-member countries in the process of economic development; and
— to contribute to the expansion of world trade on a multilateral, non-discriminatory basis in accordance with international obligations.

The original Member countries of the OECD are Austria, Belgium, Canada, Denmark, France, Germany, Greece, Iceland, Ireland, Italy, Luxembourg, the Netherlands, Norway, Portugal, Spain, Sweden, Switzerland, Turkey, the United Kingdom and the United States. The following countries became Members subsequently through accession at the dates indicated hereafter: Japan (28th April 1964), Finland (28th January 1969), Australia (7th June 1971), New Zealand (29th May 1973) and Mexico (18th May 1994). The Commission of the European Communities takes part in the work of the OECD (Article 13 of the OECD Convention).

Publié en français sous le titre :

SYLVICULTURE, AGRICULTURE ET ENVIRONNEMENT

FOREWORD

The Spanish Ministry of Agriculture, Fisheries and Food hosted an OECD Workshop on forestry, agriculture and the environment in Madrid on 17-20 October 1994. Government officials from forestry, agriculture and environment ministries from 20 OECD countries, as well as representatives from four Central and Eastern European countries, international organisations and professional farmers' organisations discussed the policy issues relating to farm forestry in OECD countries with regard to their linkages with agricultural policy reform, the environment, and possible future developments in markets for timber.

The Joint Working Party of the Committee for Agriculture and the Environment Policy Committee discussed the conclusions of the Workshop in December 1994. The Committee for Agriculture and the Environment Policy Committee agreed to recommend the derestriction of this document during their meetings in April and May 1995, respectively, under the responsibility of the Secretary-General.

The thematic papers prepared by consultants, an overview paper from the OECD Secretariat and official statements together provided a rich collection of information, experience, and ideas for the participants. A summary and assessment of the papers and discussions in the Workshop is also included in this publication.

Twenty-four countries prepared country case studies outlining their policies and experiences relating to farm forestry. The full set of country case studies is available separately as a general distribution document directly from the Directorate for Food, Agriculture and Fisheries, OECD, Paris. A list of the country case studies is included in the Table of Contents of this publication which is published on the responsibility of the Secretary-General of the OECD.

* * * *

The OECD expresses its appreciation to the countries that contributed to the financing of the Workshop on forestry, agriculture and the environment (the Commission of the European Communities, Denmark, Finland, Japan, Portugal, Spain, and the United States), and especially to the Spanish authorities for providing the major share of the financing, and for hosting the Workshop.

TABLE OF CONTENTS

Conclusions of the workshop on forestry, agriculture
    and the environment 7

Overview and main policy issues: An OECD perspective
    Gérard Bonnis, Administrator,
    Directorate for Food, Agriculture and Fisheries, OECD, Paris 13

Farm forestry, agricultural policy reform and the environment:
    a summary and an assessment of the workshop
    Peter H. Pearse, Professor
    Department of Forest Resources Management
    University of British Columbia, Vancouver 25

Economic implications of sustainable agro-forestry systems
    John A. Miranowski, Director,
    Resources and Technology Division
    Economic Research Service, USDA, Washington 45

Agricultural policy reform and farm forestry
    John Robert Crabtree, Head,
    Environmental and Socio-economics Group
    Macaulay Land Use Research Institute, Aberdeen 55

The outlook for farm forestry and the markets for its products
    UN-ECE/FAO Agriculture and Timber Division, Geneva 81

Farm forestry and the environment
    Alberto Madrigal Collazo, Professor
    Technical High School of Forest Engineers, Madrid 135

Official Statements

    His Excellency Luis Maria Atienza Serna
    Spanish Minister of Agriculture, Fisheries and Food, Madrid 159

    Gérard Viatte
    Director for Food, Agriculture and Fisheries, OECD, Paris 165

Georges Touzet
Director General of the French National Forest Agency, Paris　173

Mr. Carlos Tió Saralegui
Secretary General for Agrarian Structures,
Spanish Ministry of Agriculture, Fisheries and Food, Madrid　179

List of participants　185

* * * *

COUNTRY CASE STUDIES

Available as a separate general distribution document
from the Directorate for Food, Agriculture and Fisheries, OECD, Paris

Australia
Austria
Belgium
Commission of the European Communities
Czech Republic
Denmark
Finland
France
Greece
Hungary
Ireland
Italy
Japan
Mexico
The Netherlands
New Zealand
Poland
Portugal
Slovak Republic
Spain
Switzerland
Turkey
United Kingdom
United States

CONCLUSIONS OF THE WORKSHOP

ON

FORESTRY, AGRICULTURE AND THE ENVIRONMENT

CONTEXT

Agricultural policy reform is underway in several countries, including the European Union, at different speeds and from different starting points, and has been given added impetus by the GATT Uruguay Round agreement. In OECD economies, the agricultural sectors have experienced significant structural changes over recent decades due to technological, economic, and social changes; one of the effects has been the increasing diversity of sources of farm household incomes. While policy reform will add extra pressure on farm sectors to adjust in many OECD countries, with implications for farm incomes, labour and land use, including rural development consequences, it also provides opportunities for countries to establish an economically viable agriculture on a long term basis, which will also offer the possibility to contribute to achieving the sustainable development goals as committed at the Rio Summit.

Farm forestry may provide one of these opportunities to farmers to adjust to this structural adjustment pressure. There is also increasing awareness of the contribution to sustainable land use and to the environmental benefits associated with farm forestry. More generally, the potential contribution of farm forestry to balanced development of rural areas, countryside management, and the conservation and enhancement of landscapes, is an important objective in many countries. For all these reasons, there is an interest by some farmers and governments for farm forestry, where some farm resources (land, labour and capital) are used for forestry activities, either on-farm or off-farm.

MAIN ISSUES

The Workshop was convened to share country experiences in developing farm forestry. It was the first time in which forestry, agriculture and the environment had been explicitly considered together in the OECD. The discussions in the Workshop attempted to establish the scope of the linkages or "bridges" between forestry, agriculture and the environment and to explore policy approaches in a pragmatic and forward-looking way.

The main topics discussed at the Workshop were the role for farm forestry in agricultural policy reform; how farm forestry could achieve agricultural and environmental objectives on a sustainable basis; the possible role of specific policies; and the choice of appropriate policy measures.

On the basis of consultant reports, three panel discussions considered specific aspects of this question:

-- the impact of agricultural policy reform on the use of land and labour for forestry;

-- the impact of markets for forest products on farm forestry activities and policies; and

-- the ways that farm forestry activities and policies can best contribute to achieving environmentally sustainable land use and other environmental benefits.

MAIN CONCLUSIONS

The 24 country case studies revealed that farm forestry, like other forestry, can fulfil many functions: for the production of timber and other commercial forest products; to generate environmental benefits; and to contribute to rural development, country-side management, the conservation and enhancement of landscape, and opportunities for recreation. However, there are differences among OECD countries in the relative importance of each of these functions, and in the policy measures implemented to fulfil them.   These differences in turn spring from different agricultural, environmental and economic conditions, and different perceptions of the role of farm forestry in rural development and agricultural activities. In some countries and regions there is a long-standing integration of forestry activities on farms; in other countries there are few direct links between agricultural and forestry activities; and in some countries there are specific programmes to encourage more forestry activities on farms.

The discussion focused in particular on the multiple roles of farm forestry -- in the context of agricultural policy reform -- in the provision of timber, environmental benefits, and rural development.

It was generally felt that farmers' decisions about production of <u>timber and other commercial forest products</u> should be determined by market conditions, both globally and locally. It is necessary to look carefully at the conditions under which farm forestry could contribute to farm incomes, following agricultural policy reform. In most cases, where agricultural policy reform leads to reduced agricultural production and land values, timber production will become a more attractive option for farmers. However, farm forestry is not likely to provide sufficient income from timber and other forest output in the short term to offset any reduced income from agriculture. Moreover, agricultural policy reform may lead to extensification of farming, rather than wood production, although there may also be in the future opportunities to develop wood production for biomass.

Farm forests can play a useful role in conserving natural resources and providing net <u>environmental benefits</u>. Trees can enrich biodiversity and restore damage created by certain agricultural practices, provide a carbon sink, and play a part in arresting soil erosion and maintaining water quality. The internalisation of these external public benefits to farmers over and above private benefits can justify some policy intervention.   Although there are serious valuation problems associated with measuring all externalities, some are more difficult to quantify than others. Nevertheless, some effort needs to be made to identify the importance and distribution of these benefits. These should help in defining the nature and magnitude of any policy response.

Farm forestry, including the downstream industries, can also contribute to a variety of <u>rural development objectives</u> (e.g. employment, incomes, countryside management, landscape). It should therefore be considered as one element in broader rural development strategies;

It will be necessary to clearly define the objectives of policy and the criteria by which policy measures are chosen and designed, and to regularly evaluate and monitor the effectiveness of farm forestry policies, within a coherent institutional framework. In particular, it was emphasised that any domestic policy measures should be as least trade distorting as possible.

It was stressed that any policy responses should be clearly targeted to the respective objectives that farm forestry is intended to achieve. For example, government intervention to support farm forestry activities should only be considered where it can be demonstrated that market mechanisms do not exist or are unable to achieve the desired objectives, and any particular government intervention is shown to be the best way to accomplish the objective; information, training and education programmes may be necessary to introduce farmers to forestry opportunities, as well as to the appropriate technologies; in some cases the best policy solution may be to address the obstacles to the efficient functioning of markets and the best use of resources, including land. All of these policy responses will require an active dialogue with the farming community.

IMPLICATIONS FOR FUTURE WORK

Following the discussions in the Workshop, there is a need for further work in analysing the policy linkages between forestry, agriculture and the environment, and the evaluation of farm forestry policies. The Workshop suggested that OECD could, taking into account ongoing work in other bodies, deepen its analysis of the linkages between farm forestry on the one hand, and the production of timber, the provision of environmental goods, and rural development, on the other hand, in the context of agricultural policy reform.

In particular, work is needed to quantify the positive and negative external effects of farm forestry activities, identifying ways to overcome obstacles to the efficient allocation of resources for agriculture and forestry, and to explore the institutional arrangements to achieve a coherent approach to forestry, agriculture and environment policies. This work could be undertaken by the competent national and international bodies, paying particular attention to avoid the duplication of activities. The competent OECD Committees for their part might discuss how they can best contribute to this work, on the basis of the conclusions of the Workshop, and how they could contribute in particular to the monitoring of developments in the field of farm forestry.

OVERVIEW AND MAIN POLICY ISSUES:

AN OECD PERSPECTIVE

Gérard Bonnis, Administrator
Directorate for Food, Agriculture and Fisheries
OECD, Paris

INTRODUCTION

OECD countries have started to reform their agricultural policies and there is a renewed interest in forestry as an option for farmers in coping with structural adjustment in agriculture. Forests play crucial roles in economic and ecological systems, in erosion control, watershed management, the enhancement of biodiversity, global climate change, as well as in wood production and the provision of recreation facilities. In many OECD countries, a better management of forest resources and the reforestation of some agricultural lands could play an important role in enhancing the environmental benefits from the use of agricultural resources, while providing an alternative or supplementary sources of income for farmers.

The purpose of the present Workshop is to share experiences in Member countries, to define the key policy issues, and to suggest areas that might be considered in the future, especially with regard to the current work of the Organisation on the analysis of linkages between agricultural and environmental policies, in particular in the context of agricultural policy reform. In analysing the interface between forestry, agriculture and the environment, it would be an ambitious task for the Workshop to consider the dual-sided relations between forestry and agriculture on one hand, and forestry and the environment on the other. In order to somewhat narrow the scope of the Workshop, the following issues are considered the most important:

-- Is there a role for policy in promoting sustainable farm forestry in the context of agricultural policy reform and structural adjustment?

-- If so, which are the most appropriate policy approaches and measures?

-- How do policies to address farm forestry differ among OECD countries?

-- What would be the environmental consequences of the expansion of farm forestry?

There is a need for clear definitions and terminology used in the interface between forestry, agriculture and the environment to provide a common basis for analysis and assessement. For the purpose of the Workshop, it is suggested that the expression "*farm forestry*" refer to the orientation or use of farm resources (especially land and labour) towards forestry activities. This would include the involvement of farmers in tree planting activities on the farm, either using agroforestry techniques or in the form of woodlots, and in the management of existing farm woods to diversify their business, to improve the appearance and often the capital value of the farm, while contributing to the protection of the natural environment and the enhancement of the farm's sporting value. It might also include forestry activities realised on the farmland but not involving farm labour directly (for example, on-farm tree planting realised by public or private enterprises contracted by farmers).

Farm forestry can also include the use of farm resources for forestry activities off the farm. This embraces the participation of farmers in state, communal or private forestry activities, that allow them to remain in their local communities. The Workshop could also consider the case of forestry operations undertaken by State or private firms on that agricultural land which does not belong to a farmer or on which the tenant is not a farmer. For instance, this can be the case for the industrial plantation of fast-growing species on land released by agriculture several years ago. Farm forestry should thus not be seen only as "commercially viable wood production on cleared agricultural land", but also include farmers' attitudes to forestry and the factors influencing their decision to convert part or all of their farm resources to forestry activities.

AGRICULTURAL POLICY REFORM AND FARM FORESTRY

In most OECD countries, the agricultural sector has been characterised by high levels of support due to policies which masked the role of markets in the allocation of resources. Price support policies have accounted for the largest share of support to agriculture in most Member countries. These policies, which have encouraged greater use of capital and purchased inputs, and technological progress have contributed to a marked rise in the agricultural productivity of farm labour and land. However, commodity-related assistance to the agricultural sector has led to significant budgetary outlays and agricultural commodity surpluses, leading to a misallocation of resources, overall economic inefficiencies and to acute trade disputes. Despite support, the main features of agriculture in the OECD have been a continuous decline in farm employment and in the overall contribution of agriculture to GDP, associated sometimes with a decrease in farmers' revenues, together with an increase in the share of off-farm household income. This situation has led governments to reform their agricultural policies and to search for more suitable ways to achieve policy objectives.

The main thrust of the drive for agricultural policy reform, as agreed by the 1987 OECD Council at Ministerial level, is to let market signals influence the orientation of agricultural production, through a progressive reduction of support and protection of the agricultural sector. While pursuing this long-term objective, consideration may also be given to other concerns, such as food security, environmental protection or employment which are not exclusively economic. Rather than being provided through price guarantees or other measures directly linked to production or to factors of production, farm income support should be sought, as appropriate, through direct payments. These principles were reaffirmed in 1992 by the meeting at Ministeriel level of the Committee for Agriculture of the OECD, which also stressed the growing importance of the mutual relationships between agriculture and forestry on the one hand, and the environment on the other hand.

After the successful conclusion of the Uruguay Round in December 1993, the OECD Committee for Agriculture met at High Level in February 1994 and expressed the need to further explore the role of the various policy instruments and to define their most appropriate combination, or "policy mix", in advancing the reform process. It was recognised that, whilst market-oriented reform is generally expected to bring environmental benefits, specific measures targeted to desired environmental outcomes will also be necessary. Furthermore, the adjustment of the agricultural sector will be facilitated if it is supported by comprehensive policies for the development of various activities in rural areas, notably to help farmers to find supplementary or alternative income.

As the process of policy reform involves an overall reduction in support, it may, as a consequence, lead to changes in relative prices between commodities, regions and countries, farm inputs and outputs. In their decision on how to allocate their resources, farmers will increasingly face the changed incentives offered by markets, as well as a different set of risks and uncertainties. This will result in changes in the levels, composition and location of production, and in farm practices, and will likely have consequences for land prices. However, is should be noted that in countries (or for commodities) with very low existing levels of support, there could be an expansion of agricultural production, as these countries (or producers of these commodities) might be in a better position to exploit their comparative advantage in response to changes in market prices.

The long period of relatively high support prices in agriculture in many OECD countries contributed to deforestation for the farming of land marginal for agriculture. The reduction of administered prices will increase farmers' exposure to market forces and should contribute to the reversion of some agricultural land to forest land. Existing supply control measures also influence the allocation of agricultural resources between agriculture and forestry by establishing "rights to produce" in the form of quantitative restrictions on output (such as milk quotas) or on input use (via acreage reduction programmes, production licences or land set-aside schemes). The reduction of border protection in agriculture might also lead some farmers to increase their production of forest products for which trade restrictions are generally much more limited, and which could become commercially attractive options.

Direct payments to farmers have become more and more important in recent years in most OECD Member countries. They consist of all budgetary outlays transferred directly to producers, based on or linked to farm output (including deficiency payments) or on the production factors (for example, acreage and headage payments). While direct payments still represent a small share of the total assistance to agricultural producers (the OECD average was around 17 per cent in 1993), those payments which are not directly linked to commodity-specific production, but are farm-linked, are progressively gaining importance. Direct payments also exist in the forestry sector where they are often linked to tree planting and maintenance activities. In the context of the Uruguay Round agreement, payments specifically granted for farm forestry activities could constitute a determining factor for farmers wishing to convert part of their land into forests, especially in less favoured areas where direct payments often contribute a significant share of farmers' income.

Other forms of support of indirect benefit to agriculture are widely used in OECD countries, but do not account for a major share of overall agricultural support. These include government explicit or implicit subsidies on purchased inputs, farm credit, agricultural research and development, extension services, education and agricultural infrastructure. They also include sub-national assistance measures. These measures can also play an important role in the development of farm forestry, for instance through the reduction of input costs in the form of capital grants and tax concessions.

The decision of farmers to shift resources from agriculture to forestry will partly depend on the relative prices of agricultural and forestry products, including the relative importance of support and protection in agriculture and forestry. It will also be influenced by the longer-term perspective linked to the use of resources for forests compared with agriculture: given the time necessary for the harvesting of wood from forest tree plantations compared to other agricultural commodities, forestry generally allows much less flexibility in crop rotation than agriculture once the decision to convert is taken.

In the light of the above, the following key concerns might be addressed by the Workshop:

-- Will forestry be an attractive option in the post-reform period for farmers?

-- Should farm forestry schemes be considered as part of the overall agricultural policy reform process?

-- Are these schemes compatible with a market-led approach?

-- Should forestry be considered as one means for farmers to continue a productive-type activity on land released from agriculture while contributing to improving the environment, instead of leaving this land idle, even if this involves some assistance from the government?

ENVIRONMENTAL POLICY AND FARM FORESTRY

Forestry activities create both positive and negative environmental externalities. On the positive side, forests improve the climate by increasing the precipitation volume and stabilizing temperatures, they help to clean the atmosphere by acting as a pollutant filter (up to certain tolerance levels), and, during their growing phase, they store carbon. Forests play a vital role in protecting fragile soils, notably by controlling wind and rainwater erosion. They also play a very important function in regulating water flow, thus reducing risks of flood or drought, and in improving water quality. Forests also provide a large number of ecological niches, sheltering plants or animals. They often provide attractive scenery and bear a high symbolic and recreational values.

On the negative side, soil and water may be acidified by the litter of certain species, and intensive plantations may create hostile reactions from those in the community who closely identify forests with nature. The amount of fertilizers and pesticides used is generally much more limited in forestry than in agriculture, and their effects on soil and water pollution much less important, since they are not applied every year. However, the impact of forestry activities on the environment comes closer to agriculture in the case of intensive plantations of fast-growing species with short rotations.

In recent years in OECD countries many new policy measures have been introduced in the pursuit of environmental protection. These are sometimes sector-specific or of a more general nature but may have a bearing on sectoral policies, including those in the agricultural and forestry sectors. These policy measures can be broadly grouped according to three categories: regulatory measures -- which remain predominant, especially in the forestry sector; economic instruments providing monetary incentives or disincentives -- which often constitute a complement to regulations; and voluntary and information approaches -- which are wide-ranging, but on which the extent of actual implementation is often unclear.

The *regulatory approach* requires farmers to undertake specific obligations, or not to implement certain activities, under penalty of fines or punishments in case of violation of environmental regulations. The potential of regulations is considerable. They are widely employed in the forestry sector, mainly to protect forest resources, whether public or private, natural or planted, from depletion due to uncontrolled harvesting and over-exploitation. However, forestry regulations are often directed at wood production, with less attention paid to the various other goods and services provided by forests, including non-wood products and environmental benefits. Furthermore, if they are to be effective, they require appropriate enforcement. In meeting the requirements of regulations designed for the forestry sector, farmers may incur additional costs, which can constitute a disincentive to develop farm forestry. Moreover, they can be inflexible, and might not reflect the diversity of situations among farm forestry schemes. Delineation of zones for commercial wood production on private lands might be appropriate in some instances, especially to introduce an "as of right" use for tree planting and subsequent harvesting within the zoning framework.

The aim of *economic instruments* is to influence the behaviour of farmers or forest enterprises towards the protection or promotion of the environment, either through financial incentives or disincentives. This takes the form of a monetary transfer between the various agents: polluters, or benefactors, and the government. State budgetary constraints and the implications for administrative costs of the need to strengthen the institutional framework, are important considerations when determining the optimal rate for taxes or subsidies.

In order to take account of the environmental effects resulting from forestry activities that are not internalised or accounted for by the sector and thus not reflected in prices, farm forestry activities could be subject to taxes or charges (in the case of negative externalities), or could benefit from payments or special credit loans linked to the provision of positive environmental services. In the second case, payments might also take the form of tax concessions, as forestry is already subject to various forms of taxation, such as income or corporation tax on woodlands managed on a commercial basis with a view to the realisation of profits; capital gains tax linked to the transfer of ownership; or, inheritance tax. Various schemes of deductibility exist, for instance to cover the expenditure for the cost of planting and maintaining woodlands, and tax relief might be authorised subject to suitable undertakings being given to conserve the qualifying interest. The use of economic instruments may also be envisaged in accordance with the cultivation methods employed. In this case, conditional payments might be introduced where the right to receive a payment would be conditional upon meeting agreed criteria of sustainability in forest plantation and management techniques.

Some of the functions of forests might be considered as public goods. This would require evaluating the utility of forests for a given use, for example in the case of informal recreation and defining the optimal quantity and quality of forests to be produced. Where markets do not exist for such public goods and, as such, consumers do not pay for their consumption, their cost of provision and maintenance would need to be met by the population as a whole through the intermediary of public authorities.

Another possibility of addressing environmental concerns is through *voluntary and information approaches*. These may include agreements reached voluntarily between the producer (either causing pollution or creating benefits), and the recipient (either suffering pollution or receiving benefits). This might be achieved through direct consultations and discussions between the contracting parties, assuming relatively low transaction costs and sufficient awareness from the recipient of the production and processes methods of the producer. The prerequisite for this kind of approach is a well-defined and updated legal framework (environmental standards, property rights) to serve as a basis for any agreement. In addition, government assistance to research and development, and to the collection and diffusion of information in general, would certainly contribute to the spread of voluntary or co-operative actions of this kind.

THE KEY POLICY ISSUES ASSOCIATED WITH FARM FORESTRY

The range of objectives of farm forestry

As with agricultural activities, the role of forests is multifaceted and it is not always easy to dissociate the production of timber from the provision of environmental goods and services. Foresters always have to consider trade-offs between wood production and the environmental functions of forests, according to ecological criteria (tree species, site sensitivity) and socio-economic parameters (markets, policy concerns). However, from a policy perspective, it is important to evaluate the implications on the agricultural and the forestry sectors, and for the environment, of the objectives specific to farm forestry activities.

If farm forestry activities were considered mainly as a supplier of wood for the market, farmers would need to take account of the relative profitability between timber production and agricultural activities, before envisaging any conversion of land to forestry. In this case, it is difficult to see a role for policy intervention unless there are market imperfections. On the other hand, if farm forestry activities were considered by policy-makers as a means to protect or enhance the environment, farmers would have to decide whether or not they would be willing to set aside part of their agricultural land for a period of time for forestry activities. Here, it might be more difficult to see any shift in land use occuring in the absence of some policy intervention.

Policy decisions might be facilitated by separating the objectives of farm forestry according to the type of forest. Planted (man-made) forests might be essentially aimed at wood production whilst natural forests would mostly provide environmental services. For instance, New-Zealand relies almost entirely on planted forests for all wood needs, while any wood from natural forests that is milled has to come from a sustainably managed resource. However, young forest plantations may prove more efficient in capturing carbon than a sustainable management of old forests. The decision on whether priority should be given to the sustainable management of existing woods and forests or to a steady expansion of tree planting will thus depend on the range of farm forestry policy objectives.

Factors influencing farm forestry

Relative prices, technology, demography, consumer preferences, and incomes are among the key influences determining forestry as an option for farmers. The development of farm forestry is also influenced by regulations, particularly those related to forest management plans and those affecting changes in land use. The size of farms, the fragmentation of farming, group farming, part-time farming, the closeness to urban areas, the threshold of farm viability, the age of farmers and the type of tenure, and having a successor or not, are all structural factors that also impact on the possible transfer of farm resources to forestry uses.

The size of farm woodlands and the long-term return of forest investment are also crucial factors to consider when looking for the many diverse economic and environmental benefits that forests may provide. Farmers engaging in forestry are aware of the need to cope with the risks specific to long-term activities. To increase the short-term profitability or ensure a regular income from farm woodlots, fast-growing species might be preferred, the area of farm woodlots under management might be increased (possibly through the grouping of owners), or farmers might be paid a rent by forest investors on the land their trees occupy. Policy makers need to explore ways to change the perception of tree plantations being a long-term investment whereby capital is locked up for decades.

Policy issues in farm forestry

Policy issues associated with farm forestry are closely related to developments in trends in the agricultural and forestry sectors, in their degree of integration, and in their respective policies. The basic policy issue is between the role that policies (and which ones) and markets can play in the area of farm forestry.

In this context, it is important that the Workshop discuss the possible roles of policy and markets, by adopting a pragmatic and forward-looking approach drawing on the diverse experiences in OECD Member countries, and the material prepared for the three panels. In the suggested questions outlined below, a central theme and a checklist have been identified to help focus the general discussion on the *Country Case Studies* and on the *Panels*.

Country case studies

What is the experience of policies to promote environmentally sustainable farm forestry in OECD countries, and which are the policy instruments used to achieve this?

-- What are the perceptions of "farm forestry" and the relationship between forestry, agriculture and the environment in different OECD countries?

-- What are the policy approaches and programmes to encourage farm forestry for timber and/or to enhance environmental benefits in the context of the reform of agricultural policies in OECD countries?

-- What kind of instruments or mix of instruments are being used to encourage farm forestry, given the long rotation of forestry production (e.g. investment aids, fiscal incentives, direct payments), and how are the programmes assessed and monitored?

-- To what extent are farmers and foresters involved in the policy processes concerned with farm forestry; what has been the evolution in the attitude of farmers towards forestry activities?

-- How do policy-makers attempt to ensure a coherent approach between forestry and agricultural policies, between forestry and environmental policies, and between forestry, land-use and rural policies, while allowing market signals to guide the use of agricultural and forestry resources?

-- Given the different degrees of policy intervention in OECD and non-OECD countries in the forestry sector, what are the implications of forestry policies for international trade in wood and timber products?

21

Panel on agricultural policy reform and farm forestry

This panel will try to evaluate the extent to which farm forestry might be an option in agricultural policy reform. This is particularly topical after the conclusion of the Uruguay Round and, in particular, the commitment of contracting countries to undertake a 20 per cent reduction of domestic support to agriculture and the liberalisation of trade by the year 2000, with the consequences for reducing market price support and the level of production of agricultural commodities in some cases. This could be of particular importance in determining land use and farmers employment in forestry as part of the reform of agricultural policy.

What will be the impact of agricultural policy reform on the availability and use of land and labour for forestry?

-- Under which circumstances can forestry be a viable option for farmers or other economic agents and farmland in OECD countries following agricultural policy reform?

-- Which policies could be implemented to ensure that land and labour resources are allocated between agriculture and forestry on a market-led basis?

-- To what extent can farm forestry (as compared to other land uses such as leisure parks, wilderness, hunting areas, for example) contribute to the structural adjustment process following agricultural policy reform?

-- To what extent are the legal provisions relating to changes in the use of agricultural land (including set-aside) and permanence of forestry as a user of land, impediments to making forestry more attractive to landowners?

-- What might be the impact of farm forestry activities on the development of the economy of rural communities?

Panel on the outlook for farm forestry and the market for its products

This panel will focus on the role of farm forestry in the context of the markets and policies to orient the production of timber and other forestry products, including non-wood products. Forestry policy continues to evolve in response to changing public perceptions and increasing emphasis is placed on the multipurpose management of woodlands, taking into account economic, social and environmental considerations. For an overall evaluation, the private return from forestry investment needs to be assessed in conjunction with the expected social returns from activities aimed at the preservation of the environment, including natural habitats, landscape, or countryside assets. Given the policy measures in place in many OECD countries to promote the expansion of farm forestry, it is important to assess how the possible spread of farm forestry will impact on the markets for timber.

What will be the impact of farm forestry activities and policies on the markets for forest products in the future?

-- What are the main long-term trends for the markets of the various assortments of forest products and how could these trends be affected by increased production from farm forestry?

-- To what extent is the nature of forestry, including farm forestry, such that market signals will not lead to appropriate forestry investment decisions?

-- Are there any conditions under which farm forestry could be viable without public subsidies?

-- To what extent might there be a role for policy to ensure an economically viable size and effective management of farm woodlands?

-- What is the role for private sector investment in forestry, and how does the policy framework affect domestic and international flows of private sector investment into forestry?

-- What types of wood-using industries might be the best commercial partners of farm forestry, and is there a role for policy to support their development?

Panel on farm forestry and the environment

This panel will deal with another key policy issue of farm forestry, namely its role within broader environmental concerns. The need to protect forests from uncontrolled harvesting has been recognised since early times but new policy issues are emerging in the OECD area, including the sustainable management of forests, the conservation of biodiversity of forests, the role of forests in preventing soil erosion, the adaptation of forests to climatic changes, and the fact that the recreation value of a forest may, in some cases, exceed its timber value. More emphasis could be given to the provision of positive externalities and public goods by forests, including in farm forestry policy, but the problem of the valuation of the public good element of forests can be large.

In what ways can farm forestry activities and policies best contribute to achieving environmentally sustainable forestry and other environmental benefits?

-- What are the specific positive and negative environmental impacts of forestry and how might an expansion of farm forestry activities affect these impacts?

-- What is likely to be overall environmental impact of a shift from agriculture to farm forestry activities, including on landscape, at the urban-rural interface and on farm tourism?

-- In what ways could farm forestry be publicly assisted to provide public goods and services without causing distortion in the markets for forest products and services?

-- If policy measures can be justified to achieve the environmental benefits from farm forestry, which measures might be the most appropriate, and how should such measures take account of any environmental damage from farm forestry activities, in accordance with the Polluter Pays Principle?

-- How can policy makers best ensure a neutral policy stance between agroforestry and agricultural production?

CONCLUSIONS

It would be extremely helpful if the conclusions of the Workshop could contribute to have a better idea of the role that farm forestry could play in relation to agricultural adjustment, land use, rural development, timber production and the protection of the environment. It is of great interest to establish if there is a role for policies to encourage farm forestry in OECD countries and, if so, which are the best policy measures or approaches to adopt. It is particularly relevant to consider the possible role of farm forestry in facilitating the process of agricultural policy reform, with reference to the need for a more coherent approach between agricultural and environmental policies.

The role for OECD in further pursuing work in the area of farm forestry policies and/or in the analysis of forestry policies in relation to agriculture and the environment should also be explored by the Workshop. During its meeting held from 30th May to 1st June 1994, the Committee for Agriculture agreed that any further work in OECD on forestry issues would have to be assessed in the light of the outcome of the Workshop on forestry, agriculture and the environment. To avoid a possible duplication of activities, it would be helpful at this stage if participants could inform the Workshop of any other work of which they are aware that is being carried out on farm forestry by other international organisations. Depending on the findings of the Workshop, further work on farm forestry might be developed in the Organisation within the framework of its draft Programme of Work for 1995-1996, under the activity relating to issues in sustainable agriculture. The overall objective of this latter activity is to identify the key policy issues and approaches in this area, with a view to developing a set of common elements in the concept and practice of promoting sustainable agricultural systems in OECD countries.

The conclusions of the Workshop will attempt to draw together the results of the discussions in the various sessions, and might in addition focus on the following key policy questions:

-- Is there a need for policy measures to expand farm forestry in the OECD area, and on which grounds?

-- Drawing on the lessons learned from the experience of agricultural policy in OECD countries, how can any policy intervention to promote farm forestry be designed to achieve the intended result with least trade and economic distortions?

-- What are the priorities for further work in OECD in examining the policy linkages between forestry, agriculture and the environment and the implications for land use and rural development?

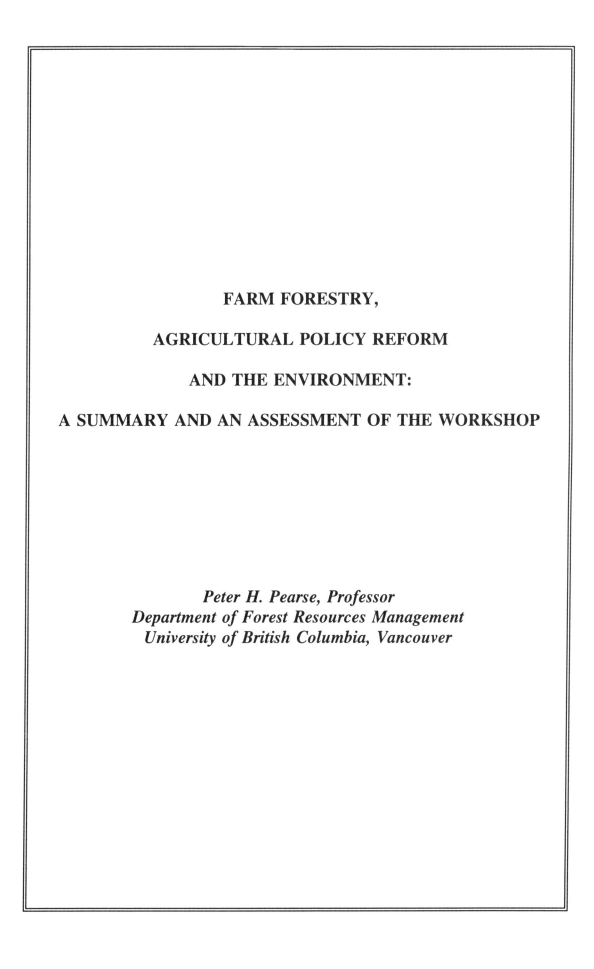

FARM FORESTRY,

AGRICULTURAL POLICY REFORM

AND THE ENVIRONMENT:

A SUMMARY AND AN ASSESSMENT OF THE WORKSHOP

Peter H. Pearse, Professor
Department of Forest Resources Management
University of British Columbia, Vancouver

INTRODUCTION

Throughout the developed world powerful forces are reshaping agriculture, farming communities and rural landscapes. Some have been at work for a long time: new technologies have been changing farming practices, raising productivity and replacing labour with capital. In most countries these and related economic trends are reflected in consolidation of farms, declining rural populations and changing landscapes. Over recent decades governments, also, have become a major influence in agricultural development, especially through programmes of assistance to farmers and regulation of markets for farm products.

Recent decisions by many OECD governments to reform agricultural policies can be expected to put new pressures on farming, and to change the pattern of agricultural development. In particular, commitments made under the Uruguay Round of the General Agreement on Tariffs and Trade (GATT) will lead to reduced export subsidies and freer international trade in farm products, with consequences for the level of domestic support.

In general, reduced support for farm production could lead to lower farm incomes and employment, and this has led some governments to search for other economic opportunities for farmers, either within the agriculture and agrofood sector, or in other sectors. One of the opportunities is to develop farm forestry, which converges with growing environmental concerns, and recognition of the range of environmental benefits that can flow from afforestation. In short, development of farm forestry is seen as a possible means of advancing policies aimed not only at agricultural adjustment and rural development but also at environmental enhancement.

This paper attempts to draw together the main threads that emerged from the documentation and discussions in the Workshop, especially as they relate to policy developments. It does not purport to summarise the consultants' reports, which appear in full in this volume, or the country statements, which are available in a separate publication. It should be regarded, rather, as the author's views on the policy issues relating to farm forestry in the context of structural adjustment, the development of the rural economy, and the environmental consequences, drawing freely upon the documentation and discussions at the Workshop.

The new interest in farm forestry raises three basic questions for governments:

1. What role, if any, can farm forestry play in policies relating to structural adjustment, the development of the rural economy and protection of the environment?

2. If there is a potential role for farm forestry in agricultural policy reform, and would it require government intervention?

3. If government intervention can be justified, which policy measures would be most appropriate?

This progression of issues, elaborated in more detail in the overview chapter prepared by the OECD Secretariat, provided a focus for the proceedings of the Workshop, as well as an organisational framework for this paper.

The paper begins, in the next section, with a brief discussion of the context -- especially the recent Uruguay Round commitments relating to trade and agriculture, -- within which governments must frame their policies relating to farm forestry. The following sections deal, in turn, with the linkages between forestry, agriculture and the environment, the market outlook for commercial forest products produced by farmers and the opportunities in farm forestry presented by the changes of agricultural policy. The discussion then turns to the rationale for government support for farm forestry and other possible forms of intervention. The paper concludes with some general observations on the range of views expressed in the Workshop.

The discussion on the following pages attempts (as did the participants at the Workshop) to take a pragmatic and forward-looking approach to the question of the scope for farm forestry and the role of government. Among OECD countries there is already a growing experience with policies and programmes for forestry as well as agriculture, and the sharing of experiences was an important purpose of the Workshop.

POLICY CONTEXT

The question of farm forestry needs to be considered within the broader context of policy objectives of governments. These vary considerably, but among OECD countries the following are the most general and relevant to the present discussion.

Commitment to markets

Throughout the developed world, there is widespread and growing commitment to markets as the means of guiding economic activity. This trend is most evident in the transition economies of Central and Eastern Europe, as reflected in the Workshop presentations by Hungary, Slovakia and Poland. But it is also growing in the already largely market-oriented economies of OECD countries as well.

The growing reliance on market forces is expected to have a significant impact on agriculture. In many countries governments, having built up elaborate programmes to promote and regulate agricultural production, marketing and international trade, now find that these policies have led to surplus production, budgetary stress and conflicts over international trade, among other problems. This was clearly recognised in the communiqué issued by Finance and Foreign Ministers following the 1987 meeting of the OECD Council, in which they identified domestic support policies as the main cause of the serious distortions in agricultural markets, and advocated reforms to allow market signals greater scope in guiding production and trade. These conclusions, and principles for reform, have been reaffirmed by subsequent OECD Councils.

All this means that any new policies that might be adopted to promote farm forestry must be consistent with the broader objective of giving market forces greater sway in guiding agricultural production and trade. As emphasized by New Zealand, the United States, Sweden, Mexico and several other countries at the Workshop, any intervention must minimise distortions of markets. In particular, forestry policies should avoid causing problems of surplus production and dependency on government assistance of the kind that have resulted from agricultural policies.

This is not to say, of course, that markets always work perfectly or give clear signals to producers. Later in this paper, externalities and other market failures affecting the production of environmental and rural development benefits are noted as potential rationales for government intervention in farm forestry.

Freer international trade

Increased reliance on market forces is accompanied by a related commitment to freer international trade. Evidence of this commitment is found in the development of large regional free-trade areas such as the European Union, the North American Free Trade Agreement and Asia-Pacific Economic Co-operation, most of which expect to expand their membership. On a broader scale, the 1994 GATT agreement in Marrakech, following eight years of negotiations in the Uruguay Round, provides a framework for general trade liberalisation. Important provisions relating to agriculture include commitments on freer access to markets, reduction in export subsidies and reform of those domestic support policies that contribute to trade distortion.

These new commitments to freer trade complement the market orientation of domestic policies, adding not only pressure but also opportunities for structural adjustment in the agricultural sector. They imply, also, that policies relating to forestry must be consistent with this new trading environment. While forest products are already traded much more freely in international markets than agricultural products, it is clear that production of commercial products from farm forests will have to respond increasingly to international supply, demand and prices. Thus it is prices in international markets (and expectations of future changes in those prices) which, coupled with costs of production, will determine the underlying financial returns in farm forestry and decisions to invest in farm forestry.

Protection of the environment

Governments of all OECD countries are concerned increasingly about environmental problems. This was reflected in the 1987 report of the United Nations Commission on the Environment and the Economy (the Brundtland report) which emphasised the link between economic development and the health of the natural environment, and promoted the idea of "sustainable development." The 1992 UN Conference on Environment and Development (UNCED) focused on this theme, and drew attention to environmental impacts of agriculture and forestry. Since then, the European Council and many national governments have begun to develop plans to implement the "Forest Principles" relating to sustainable management of forests. All OECD countries are attempting to respond to domestic environmental pressures, in addition to these international commitments. The problems and responses, however, vary widely.

The significance of this trend in the present context is that the environmental impact of forests will be important in policy decisions about farm forestry.

Moreover, in addition to its direct commercial benefits, farm forestry, integrated with livestock and other farm production, can contribute to sustainable agricultural policies. Woodlands are usually regarded as contributors of environmental benefits, ranging from improved wildlife habitat to carbon sequestration. But forestry can also cause environmental damage, such as soil erosion and loss of biodiversity resulting from inappropriate forest practices and species. Accordingly, the environmental costs and benefits of farm forestry must be weighed by policy-makers in relation to alternative uses of the land; any intervention needs to be designed to promote the beneficial effects and minimise the damaging ones. The value of those environmental benefits which yield no returns to forest owners will be important in designing sustainable development policies linking agriculture, forestry and the environment.

Rural development

Many countries, especially those of Europe and Japan, are deeply concerned about rural development and the potential contribution of farm forestry. This general interest embraces a variety of more specific concerns which differ considerably among countries. For example, Italy, Ireland and Japan, among others, emphasise the need for new employment opportunities to help stabilise rural populations and communities; the United Kingdom and Denmark, where the forest cover is low, put higher priority on the environmental value of farm forestry including its contribution towards the aesthetic appeal of rural landscapes; France views forestry as part of its broad approach to countryside management ("aménagement du territoire"), Finland and Portugal see forestry as a potential means of facilitating the shift of resources out of agriculture by providing an alternative productive use of farmland rendered surplus under agricultural policy reform; other countries stress the need to diversify and supplement farmers' incomes.

The importance of this for farm forestry policy is that governments will want such activity to be consistent with broader rural development policies, and that the nature of the contribution they will be seeking will vary among countries.

Fiscal efficiency

Many national governments are now faced with fiscal pressures, and are likely to resist policies that draw heavily on their budgets. Indeed, the budgetary burden of agricultural support has been influential in recent moves toward policy reform. In their approach to farm forestry, governments can therefore be expected to resist policies that would draw heavily on their budgets. And they can be expected to seek fiscal efficiency in any programme spending, meaning that the measures chosen are the least costly means of achieving the objective, and that funds are expended in the most effective way possible.

The pressure on government spending has been exacerbated in recent years by high unemployment in many OECD countries. In rural areas this has aggravated the long-term erosion of employment in both agriculture and forestry, and the gradual loss of political influence to urban centres. This means that policy changes in both agriculture and forestry will be viewed in terms of their employment effects as well.

* * * *

The foregoing objectives are widely shared among OECD countries, and can be expected to influence approaches to farm forestry, providing an important context for discussion about the potential role of farm forestry and appropriate government policies. The relative importance of these objectives varies among countries, of course, and each country will be concerned, as well, about other issues specific to its own circumstances.

OTHER POLICY CONSIDERATIONS

In addition to the need to reconcile farm forestry programmes with broader policy objectives, OECD governments must take account of the prevailing circumstances of potential producers and their markets. Two such factors are especially important to later discussion in this paper. One is the attitude of farmers to forestry, which has been shaped by history in different ways among countries. Forestry and farming compete for the land and other resources of rural economies, and in some countries, such as Austria, Switzerland, the Nordic countries and Japan, farmers have traditionally been involved in forestry. Not all small-scale forest enterprises involve farmers, however; even in Finland half the small forest owners are not farmers. Nevertheless, in these countries farmers generally view forestry as an adjunct to the rural economy.

In contrast, in some countries of the European Union, forestry and farming have become progressively disassociated through history, so that today farmers are typically unfamiliar with forestry. In Canada, the United States, and New Zealand, farmers have traditionally regarded forests as an obstacle to developing agriculture. Indeed, farmers in some countries have an aversion to forestry. In Ireland and Spain, at least, this prejudice is apparently rooted in the dispossession of peasants, centuries ago, to make way for woodlands sought by feudal landlords.

These differences in farmers' attitudes toward forestry are important in the present context. They can imply a role for government in providing information and technical advice, demonstration projects and education to overcome obstacles of misunderstanding and ignorance. Indeed, forestry programmes might prove to be means of forging links between agriculture and environmental protection; for this reason Italy stressed the need to better integrate forestry and farming. The scope for such policies, which call for close dialogue between governments and farmers and their organisations in programme development, varies widely from country to country.

The other consideration is that farm forestry, and farm forestry policies, are by no means new to OECD countries; most already have programmes in place to encourage or regulate this activity. Data presented in the background paper by Tim Peck show that all OECD countries have significant forests, averaging about 22 percent of their total land area but varying greatly — from 6 percent in Ireland to 67 percent in Japan and 69 percent in Sweden. The portion that can be considered to be under farm forestry, small-scale or woodlot operations linked to non-industrial, rural economies is not consistently documented, but it almost certainly varies even more widely.

It is important to note that forestry policies, as they currently exist in most OECD countries, are significantly different from agricultural policies. Although most governments are involved in forestry through ownership of forest land, production of timber and regulation of private forestry, forest products are traded in much freer markets, and there are no general price support programmes as there are for farm commodities. Moreover, the availability of substitutes for many forest products has constrained trade protection policies. Thus policy-makers seeking to develop farm forestry are faced with existing forest policies which are generally less distorting than agricultural policies, and could offer a better opportunity to target programmes to serve new objectives.

MARKET FAILURES AND THE
RATIONALE FOR GOVERNMENT INTERVENTION

The reinforced commitment to the market system by most OECD countries nevertheless recognises that markets are imperfect and do not always give clear signals; the social value of some goods and services is not always reflected in their prices, and factor markets can fail to direct resources to their highest use. Thus Spain, Portugal and France, among other countries, pointed to the value of forestry in providing environmental and rural development benefits which are not measured by market prices, and the presentation by John Miranowski emphasises the importance of "getting prices right" by including external costs and benefits.

For many countries, it is the need to correct deficiencies of markets that provides a rationale for government intervention. Where the full social value of some activity is not reflected in prices charged by producers, or has no price at all -- such as carbon fixation by tree plantations -- governments can correct or compensate for this market failure through measures to stimulate the beneficial activity. And, conversely, when an activity has adverse external effects — such as pollution of waterways by wood-processing — governments can intervene to discourage or prohibit it. To the extent that governments succeed in correcting market failures and price distortions the efficiency of the economic system as a whole is improved.

Not all countries agree that market failures justify government intervention in farm forestry, however. New Zealand, among others, maintains that there are few costs and benefits of forestry that do not accrue to the landowners, and therefore there is little scope for constructive government intervention. Other countries, such as Australia and Sweden, have recently moved toward this position, and have reduced subsidies in support of forestry.

Most participants at the Workshop agreed with the general notion that markets sometimes fail to give accurate signals, and that governments can have a role in correcting or offsetting the resulting misallocation of resources. The problem, and the source of considerable divergence of views, is the prevalence of external benefits and costs associated with farm forestry, the quantification and evaluation of them, and the best means of dealing with them. These issues are discussed below.

At the root of many of the presentations and much of the discussion at the Workshop was the fundamental policy question: how much can be left to market forces or, conversely, how much government intervention is needed to correct market failures? The Workshop participants discussed at length the various possible rationales for government support for farm forestry in the context of agricultural policy reform. These can be grouped into several categories:

. to compensate for environmental benefits;

. to compensate for rural development benefits;

. to overcome market barriers and other obstacles to the development of farm forestry;

. to facilitate adjustment to policy change;

. to increase or diversify farmers' incomes.

These rationales are mainly concerned with economic efficiency through the desirability of eliminating or reducing market distortions that otherwise prevail, but they have equity or distributional implications as well.   The first two -- payments for environmental and rural development benefits of farm forestry -- attracted most discussion at the Workshop. These benefits are extremely diverse, as noted above. Their relevance and form varies greatly from country to country, providing considerable scope for debate about appropriate government policies.

In general, the case for government support rests on the existence of significant external environmental or rural development benefits -- that is, social benefits -- that are not directly captured by producers and not entirely incidental to profit-maximising production. The diversity of such benefits and the variation in their occurrence and importance means that they must be identified and assessed in each set of circumstances and, where intervention is called for, it should be carefully targeted to deal with the particular problem at hand.

Participants at the Workshop identified a variety of other obstacles to the best use of farm resources that could justify government intervention in support of forestry in particular circumstances. These include weak markets for forest products, diseconomies of small-scale operations, capital markets that are inaccessible to farmers for long-term investments, and other market imperfections noted below. These more conventional barriers to efficient use of resources may warrant measures to stimulate competition in product and factor markets and to improve market performance in other ways.

Changes in established policies, such as agricultural policy reforms, inevitably threaten to dislocate, to some extent, established farmers and farm communities. Therefore government may play a role in facilitating adjustment by easing the transition from one policy regime to another. This calls for adjustment policies which are explicitly temporary and aimed at cushioning the social impact of change.

Finally, it is often deemed necessary to support farmers' incomes on purely distributional grounds. Assistance is thus rooted in the concern for equity in the distribution of income; more specifically the relatively low incomes in the farm sector and in rural regions.

It became clear in Workshop discussions that many countries' concern about the adequacy of farm incomes has been sharpened by the recent GATT commitments to reduce price support and export subsidies for agricultural commodities. In view of these commitments, governments may view assistance to farm forestry as part of the reorientation of support towards decoupled direct payments which can cause less distortion in product markets.

LINKAGES BETWEEN AGRICULTURE,
FORESTRY AND THE ENVIRONMENT

Forestry and agriculture are both forms of rural production, capable of utilising some of the same resources of land, labour and capital. To this extent they are alternative activities. Both have significant impacts on the environment, sometimes favourable and sometimes damaging.

It is widely recognised that farm forestry can contribute to rural development through increased employment opportunities and diversification of farm incomes, and can provide environmental benefits as well. The range of these benefits is well documented in the background paper by Madrigal Collazo. However, in order to make its fullest contribution, farm forestry must be developed in certain ways. Countries such as Japan point out that the provision of economic benefits to rural communities depend on engaging local farmers rather than outside investors in the forestry activity.

Some countries see potentially substantial environmental and rural development benefits from farm forestry: the United Kingdom, France, Spain, Portugal and most other countries of the European Union point to significant social and environmental benefits. The United Kingdom and Switzerland suggest that landscape benefits depend on diversified land use in which agriculture is mixed with forestry on a small scale, in hedgerows or in patterns that benefit wildlife and improve recreational opportunities. Denmark and Greece suggest that the forests should consist of mixed native species rather than on monocultures of exotic species. The forests and forest practices that contribute most to broad social objectives relating to rural development and the environment -- which vary from country to country and region to region -- will sometimes differ from those that are most profitable to landowners and farmers.

The environmental benefits of farm forestry are especially complex. For present purposes it is helpful to recognise that the beneficiaries vary as well. Some environmental benefits accrue mainly to the landowners: for example, farmers in the United Kingdom often value woodlands as aesthetic and sporting improvements to their farms; in New Zealand trees provide valuable shade and shelter for livestock; in Spain, Italy and Austria they stabilise soils and water levels; and in other European countries they provide owners with fuel and building materials. To the extent that the benefits accrue to the farm enterprise, they are private (not public) goods. Sometimes such benefits are realised by landowners in monetary terms — such as hunting revenues where woodlands support game, or in higher property values — but often they are not. The important point is that some environmental benefits of farm forestry accrue to the forest producers themselves. To the extent that they do, the "externalities" argument for any government intervention does not apply.

Other environmental benefits accrue to local populations generally. The beneficiaries of improvements to landscapes and woodland recreational opportunities, and of watershed protection, typically include all the residents of the region, and often visitors as well. Still other benefits accrue to whole nations, such as the contribution of woodlands to national targets for green space and protected areas, or to objectives relating to biodiversity or sustainable development. Finally, some benefits are global. For example, France, Poland and several other countries drew attention to the atmospheric benefits of sequestration of carbon in forest growth. Switzerland among others, noted the potential of wood as a renewable and environmentally friendly source of energy, and the

Netherlands suggested a possible role for farm forestry as a means of relieving demands on depleting stocks of tropical timber in developing countries.

These distinctions lead to several important observations for policy purposes. First, the environmental benefits that farm forestry can generate are very varied, as reflected in the reports from individual countries. Second, these reports also reveal enormous differences among countries in the type of environmental benefits that are important, and indeed in the importance of environmental benefits generally. There is no single environmental value (in contrast to, say, timber, which is a reasonably well-defined and consistently valued product in all countries). Environmental benefits must be regarded instead as a broad category of goods and services, only some of which are relevant at any time and place. Indeed, their relevance varies greatly within countries — within even small countries, as Belgium points out. The environmental benefits of farm forestry in each country and locality must be identified and evaluated separately in each case.

Third, some, but by no means all, environmental benefits of farm woodlands are public goods, in the sense that they accrue broadly to people who do not pay for them and their value cannot be captured by the farmers who produce them. Examples include the aesthetic enhancement of rural landscapes and protection of global biodiversity. Such public goods may offer scope for public intervention, but it is recognised that these are important practical problems involved for policy makers: the measurement of the value of these public goods; the identification of the beneficiaries of these public goods (at the local, national or even international level); and the best way to take them into account.

Finally, it must be recognised that farm forestry can produce environmental harm as well as benefits. Sometimes forest practices damage soil or wildlife habitat, poor harvesting practices may cause erosion or disrupt watersheds, plantations of exotic species can displace natural biodiversity, and some types of forestry disfigure landscapes. Like environmental benefits, environmental damages or costs vary widely in relevance among and within countries. In consequence, the balance between benefits and costs varies widely as well. In this regard, it is important that the Polluter Pays Principle (PPP) be applied. Thus, when discussing the need for government intervention, the external costs of farm forestry activities need to be taken into account as well as any external benefits.

All this implies that very few generalisations about the environmental benefits of farm forestry can be made. The same is true of agriculture, of course. This is important, because it is often simplistically suggested that agriculture causes environmental problems while forestry generates environmental benefits; therefore reduction of farm subsidies, by shifting land use from agriculture to forestry, will lead to environmental improvement. The preceding discussion suggests, instead, a sequence of questions, which are dealt with in the remainder of this paper.

.   Will the agricultural policy reforms being implemented by OECD countries offer new opportunities in farm forestry?

.   How much, if any, government support for farm forestry would be necessary to make it viable, and is such support justified by its environmental and other external benefits?

.   What kind of government activities would be most effective?

MARKETS FOR FARM FOREST PRODUCTS

Primary forest products are traded relatively freely among producing and consuming countries — much more so than agricultural products — and their markets are becoming increasingly global. So expectations about future prices of forest products produced in OECD countries, including those produced on farms, should be based on expected trends in international prices.

The background paper presented by Tim Peck provides a detailed assessment of the supply, demand and price trends in world markets for forest products. FAO data shows little observable trend in the real (adjusted for inflation) prices of primary forest products over the last three decades. Indeed, the prices of products most relevant to farm forestry — pulpwood and logs — show, if anything, a slightly declining trend. Peck does not foresee changes in these trends, because the world supply of timber is being boosted by new plantation forestry and demand is being dampened by recycling of used paper and a broadening array of substitutes for wood. Low quality wood, especially, appears to be increasing in supply faster than demand in industrialised countries.

Some caution is warranted in drawing inferences about future trends from these historical data. For one thing, the statistics are broadly aggregated over world regions and over species and grades of timber, so they might disguise conflicting trends in specific markets. For another, technological developments — in generating energy from biomass, for example — might well boost future wood demand and upset recent price trends. New Zealand, at least, predicts growing demand for softwood fibre in the Pacific Region.

Although the world's forests are diminishing, this is not generally the case in OECD countries. In almost all Member countries, the area of forest is increasing, and is expected to continue to grow. Currently, some 250 000 hectares of formerly non-forest (mainly agricultural) land is being afforested annually in Europe alone, and over the past half century both the forested area and inventory of growing stock have expanded substantially. In countries such as Denmark and Ireland this expansion has recently accelerated under government support programmes. Fellings are expected to increase, but growth in increment is expected to continue to increase faster. As a result, harvests will remain at about 70 per cent of growth rates, and the growing stock will consequently increase.

Exceptions to the expansion in forested area among OECD countries are the United States and, most notably, Mexico, which has one of the world's highest rates of deforestation.

The balance between production and consumption of forest products varies greatly among countries; Canada is the world's biggest exporter, while Japan and the European Union are net importers of most categories of wood products.

Analysis of the trends in markets in forest products provides little support for predictions of higher future prices in general. However, even if real prices decline, this does not mean that forestry will necessarily become less profitable — that depends on trends in costs of production as well. And as France has emphasised, general trends may obscure profitable opportunities in local, specialised, "niche" markets.

At present, farm forestry is profitable in some countries but not in others. It is apparently economically viable without government assistance in New Zealand, Finland and Sweden, in parts of the United States and Canada, and probably, in certain circumstances, in Central and Eastern Europe, but it is rarely if ever so in Austria and Switzerland, the United Kingdom, Spain, Portugal, France and most other countries in southern Europe. In these countries, too, forestry is expanding, but it is mainly driven by government programmes of public forestry and of support for private forestry.

Several conclusions can be drawn from this analysis. One is that the profitability of farm forestry now varies greatly among countries. A second is that it would be imprudent to base new policies on the expectation that markets for forest products, in general, will strengthen to significantly change this in the foreseeable future. A third is that, in consequence, significant expansion of farm-based forest production will depend on either reduced agricultural production that will free resources for forest use, or strengthened programmes of government support.

STRUCTURAL ADJUSTMENT IN AGRICULTURE
AND OPPORTUNITIES IN FARM FORESTRY

The background paper by Robert Crabtree examines the relationship between farm forestry and reform of agricultural policies. Recent commitments to agricultural policy reform can be expected to lead to reduced support for domestic prices and export subsidies for farm commodities. But this should not be exaggerated. The Marrakech agreement calls for only 20 percent reduction in domestic price support over five years. Moreover, that reduction is likely to be accompanied by a shift toward direct income support for farmers, and payments for activities that generate environmental benefits. So it is not clear that total support for agriculture will decline significantly, if at all.

The stimulus to forestry resulting from the reduction in agricultural price support will depend on the nature of the reforms, and particularly whether they result in lower land prices and lower prices for alternative products within the agricultural and agrofood sector. For example, if price support were replaced by payments to farmers based on the area they cultivate, land prices would probably not fall significantly and incentives to convert land to forest would not result.

Institutional, fiscal and regulatory arrangements also impede the shift from farm to forest land use. In New Zealand, for example, tax disincentives to farm forestry proved to be a major barrier, and their removal in 1991 contributed to a massive conversion of farmland to forestry (and, incidentally, strengthened New Zealand's view that reduction in agricultural subsidies is an insufficient justification for intervention in other forms). In Japan, policies in support of rice production make it difficult for farmers to convert farmland to forestry. Land set-aside schemes such as those of the European Union do not encourage afforestation of land taken out of agricultural production. And where farmers receive payments for the area they cultivate, forested areas often do not qualify.

In general, if agricultural policy reform reduces support for traditional farm production, some decline in land values and improvement in the relative returns to forestry can be expected. But the stimulus to forestry might not be very great, nor sufficient to overcome barriers of unfamiliar technology, limitations of land tenure, initial capital requirements and the long waiting periods for returns on forestry investments. Agricultural reform, on its own, cannot be counted on to shift land into forestry and, in many countries at least, farm forestry probably has a limited role in agricultural reform unless substantial government support for it can be justified on grounds of external environmental or rural development benefits.

APPROACHES TO GOVERNMENT INTERVENTION
TO SUPPORT FARM FORESTRY

Where support for farm forestry is deemed necessary or desirable, the question of the appropriate form of any intervention must be addressed. The best form of government action depends on the problem to be resolved, such as the obstacles to farm forestry that need to be overcome, or the lack of compensation to woodland owners for unpriced environmental benefits they generate. Governments may be concerned with more than one objective in stimulating farm forestry, and multiple objectives inevitably complicate the design of effective programmes.

The range of possible policy measures can be ranked in order of the degree to which they intrude on market processes. The list that follows begins with measures aimed at improving markets, progressing to more intrusive forms of intervention.

Information, education and technical assistance

The efficiency of markets in allocating resources depends on the knowledge of producers and consumers about markets, products and technology. Earlier in this paper reference was made to the unfamiliarity of farmers with forestry in many countries, the lack of a forestry tradition or appreciation of opportunities in forestry among farmers, and sometimes their aversion to it, as in Ireland and Spain. These circumstances hinder appropriate market responses to alternative land uses, and may justify government programmes of information, education and technical assistance. New Zealand, for example, has adopted such measures to support the general reorientation of its economy toward markets.

Market development

A variety of imperfections in the markets for both inputs and outputs may hinder farm forestry and call for corrective government action. The most common concern about factor markets relates to farmers' access to capital for forestry purposes. The substantial initial investment required for afforestation, and the long investment period before returns are realized, exacerbate this problem. Where long-term capital markets are poorly functioning, this may call for government intervention to facilitate access to credit for farm forestry. Obstacles in markets for labour and land also exist in some countries, which may impede the functioning and developments of markets.

With respect to product markets, France and other countries have pointed to deficient markets for forest products in some regions and the need to develop farm forest production consistent with the needs of local wood manufacturing industries. Where farm producers of forest products suffer from diseconomies of small scale in product marketing, encouragement of cooperative arrangements may be warranted. Improved transport systems may also enhance market opportunities.

Removing policy obstacles

In most OECD countries, the biggest obstacles to farm forestry are agricultural policies. Such policies distort relative prices, including land values, and thereby bias land use against forestry and in favour of agricultural production. Accordingly, policy-makers seeking to encourage farm forestry should look first to the possibilities of reducing tax preferences and support for agricultural production, and of removing barriers to trade in agricultural products. Set-aside schemes and payments based on areas farmed also discourage forestry if forest products are not recognised as products qualifying for those payments.

Some policies affect international markets as well. Mexico expressed concern that, under the North American Free Trade Agreement, subsidised forest products from Canada and the United States might depress prices and incentives for afforestation in Mexico. In the United States and Canada, governments restrict export sales of unmanufactured logs, which narrows markets and depresses the prices for forest producers, particularly small unintegrated producers.

Strengthening property rights

The Workshop learned about a variety of institutional impediments to farm forestry. Among these, deficiencies of land tenure are most common. Most farmers in the OECD area own their farms, but many in Britain, Ireland and elsewhere in Europe are tenants, whose opportunities to benefit from forestry are often limited because landlords retain rights to trees on their land, or because tenants rights are insufficiently secure and long-term. In Canada and the United States, among other countries, existing legal arrangements prevent farmers engaged in forestry from benefiting from enhanced wildlife, game and fisheries. And in most European countries public rights of access to woodlands, including farm woodlands, make it difficult for owners to benefit from improving the recreational value of their forests. These obstacles could often be removed through reform of property rights systems.

In Central and Eastern European countries, more profound institutional change is taking place in the shift from centrally planned to market economies. But in other countries as well, notably New Zealand, privatisation and deregulation are being pursued as part of the increasing dependence on market forces.

Direct income support

Direct income support refers here to budgetary financed payments made to farmers, which are not linked to their production of agricultural products. Such payments can serve to supplement farmers' incomes with less distortion of product markets than subsidies tied to farm products or tax incentives for certain activities (see below). However, any such payments incur a transfer of resources within the economy, with implications for the allocation of resources.

entives for forestry

A variety of direct payments and tax incentives, familiar in agricultural policies, might be employed to encourage farm forestry. The common characteristic of these instruments is that they influence producers' voluntary decisions by changing their economic incentives.

Sometimes governments make payments to producers for goods and services that would not otherwise generate income, such as unpriced environmental benefits. Such payments are akin to a price for a product, and thus can be viewed as a correction for a market failure. Spain, Portugal and Italy, all major recipients of the EU farm forestry scheme, stress the role of such payments in generating environmental benefits.

Subsidies and tax incentives for farm forestry imply increasing farmers' financial returns from afforestation. They therefore offer means of promoting farm forestry where it is valued for non-commercial reasons and the obstacle to its development is simply a low rate of return to private investors.

Regulation and zoning

Most countries have experience with laws and regulations that restrict the way land can be used, require reforestation after logging, control logging practices and in various other ways regulate forest practices. The distinguishing characteristic of the regulatory approach is that it constrains behaviour under threat of penalties, imposes costs of compliance, and thereby may prevent forest owners from pursuing their best economic opportunities. In contrast to subsidies and taxes, which influence market incentives to which farmers can respond according to their own advantage, regulations constrain individual behaviour and to that extent prevent market signals from directing economic activity.

Nevertheless, restrictive regulations offer effective means of preventing harmful practices, and often the most expedient means of preventing adverse external impacts on others, such as pollution. Zoning provides a means of promoting certain forms of land use over others, and can also be a useful adjunct to other governmental programmes by providing a means of targeting areas that will benefit most from environmental and rural development assistance.

FURTHER OBSERVATIONS

While the discussion at the Workshop and the case studies prepared by participating countries revealed wide differences in approaches to farm forestry, there was some convergence of views about emerging opportunities and policy choices as reflected in the Workshop's conclusions. The following generalisations should be regarded as the impressions of an independent observer of the proceedings.

First, agricultural policy reform, and the general shift toward more market-driven patterns of production, can be expected to offer new opportunities in farm forestry. Reduction of support for farm commodities will influence the relative prices of agricultural and forest products in favour of the latter. However, this effect may not be sufficient to stimulate much new farm forestry especially in the short term. The effect will depend on how agricultural policies are restructured, and particularly the extent to which governments continue to support agriculture in other ways. Importantly, whether land, labour and other resources are released from agriculture (thereby becoming potentially available for forestry) will depend on the way in which new agricultural support policies are implemented, especially with respect to land withdrawal and set-aside programmes.

The profitability of farm forestry varies widely among countries, and can be expected to continue to do so. The outlook for world supply and demand for wood products suggests that decisions about farm forestry should not be based on expectations of real price increases over the foreseeable future. Significant expansion in farm-forestry will therefore depend on either increased availability of resources as a result of agricultural policy reform or on strengthened programmes of government support.

Second, it is generally agreed that farm forestry can yield environmental benefits, ranging from soil stabilisation and watershed protection to wildlife, biodiversity conservation and removal of carbon dioxide from the atmosphere. It can also provide indirect benefits in the form of substitutes for non-renewable forms of energy and for timber produced from natural forests which are not managed sustainably. Farm forestry can also generate complementary rural development benefits, by providing additional sources of income and employment, diversification of economic activity, recreational opportunities and enhancement of landscapes.

However, it is also acknowledged that farm forestry is not always the best means of providing particular environmental benefits. Indeed, it sometimes results in environmental harm; it can cause soil pollution by fertilisers, physical disturbance and other ways familiar in agriculture, and it can disfigure landscapes through choice of species, patterns of afforestation or harvesting practices (although most of these problems can be minimised with proper management).

The opportunities to generate environmental and rural development benefits through farm forestry vary enormously among countries. Within countries, also, the potential benefits vary in relevance and kind.

The difference in views among countries about the justification and desirability of government intervention to encourage farm forestry, is based not so much on the principles involved as on the empirical evidence of the external benefits. Almost no generalisations can be made about the occurrence of such benefits across OECD countries, their form, or their value. Nevertheless, the opportunities to generate environmental and rural development benefits provide the primary grounds for government support for farm forestry.

Third, because the benefits vary so much in kind and in their occurrence, there are few, if any, advocates of general subsidisation of farm forestry. Indeed, Workshop participants were sensitive to the potential danger that financial assistance to forestry would give rise to problems in forest product markets akin to those that have arisen from subsidies to agriculture -- surplus production, distorted patterns of land use and impediments to trade. The general view was that any support programmes should be as little distortive of markets as possible.

Consistent with the rejection of across the board subsidisation, there is wide agreement that any policies for farm forestry should be carefully targeted to identifiable environmental and rural development opportunities. Many countries expressed a concern not only to ensure that government support is focused on specific needs and opportunities but also that it is the least costly and least trade distorting means of achieving its purpose.

With respect to potential forms of government action, there is widest support for actions aimed at improving market signals, sharpening farmers' economic incentives, and providing good information. This includes education, technical assistance, and related measures to inform farmers of their opportunities in forestry; improved access to credit and product markets; payments for otherwise unremunerated public benefits and charges for damages inflicted on others; and other actions to help market forces to work as efficiently as possible. Such measures are consistent with the broader policy objective of increased reliance on market forces in guiding production.

Fourth, most countries already have policies and programmes relating to private forestry. These, for the most part, are directed to the particular needs and circumstances in each country, and can be built upon and modified to respond to new problems and policy objectives.

Finally, it is important to note general agreement on a fundamental point of departure on the subject of farm forestry. Forestry, like agriculture, should be regarded not as an end in itself but as a component in the integrated use of land, water and human capital. The underlying policy objective is the sustainable development of natural resources and human communities. In this role, forestry undoubtedly occupies an important place, and in many OECD countries an expanding one.

The Workshop and the supporting documentation revealed considerable scope for further investigation to assist in the design of farm forestry policies. Key issues needing clarification are the fundamental economics of farm forestry and hence its dependence on public support, the contributions that farm forestry can make to rural development and environmental improvement and, to the extent that these social benefits warrant public compensation or assistance, the forms of support for farm forestry and ways of targeting it to achieve the desired results most effectively.

BIBLIOGRAPHY

CINOTTI, B. 1992. Les agriculteurs et leurs forêts. *Revue Forestière Française.* Vol. 4.

COUNCIL OF THE EUROPEAN COMMUNITIES. 1992. Council Regulation (EEC) No. 2078/92 (On agricultural production methods compatible with the requirements of the protection of the environment and the maintenance of the countryside) *Official Journal of the European Communities.* No. L 215/85-89.

COUNCIL OF THE EUROPEAN COMMUNITIES. 1992. Council Regulation (EEC) No. 2080/92 (Instituting a community aid scheme for forestry measures in agriculture) *Official Journal of the European Communities.* No. L 215/96-99.

GOVERNMENT OF DENMARK. 1994. *Strategy for Sustainable Forest Management.* Ministry of the Environment, Copenhagen.

HORGAN, G.P. 1994. *The Restructuring of New Zealand Forestry.* New Zealand Forest Research Institute Ltd., Rotorua.

PECK, T.J. and J. DESCARGUES. 1994. The Policy Context for the Development of the Forest and Forest Products Sector in Europe (Advance version). Ecole Polytechnique Fédérale. Zurich.

UN-ECE/FAO. 1992. The Forest Resources of the Temperate Zones. *1990 Forest Resource Assessment* (2 vols.). United Nations. New York.

UNCED. 1992. *The Forest Declaration.* United Nations Conference on Environment and Development. Rio de Janeiro.

VOLZ, K.R. and N. WEBER (eds.). 1993. *Afforestation of Agricultural Land.* CEC, Brussels.

WIBE, S. 1992. *Market and Government Failures in Environmental Management. Wetlands and Forests.* OECD. Paris.

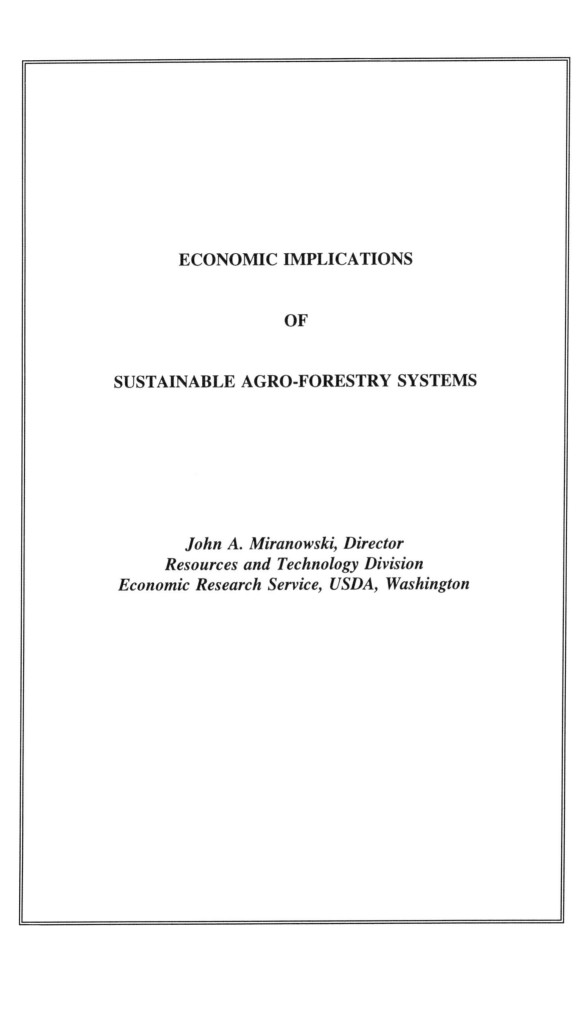

ECONOMIC IMPLICATIONS

OF

SUSTAINABLE AGRO-FORESTRY SYSTEMS

John A. Miranowski, Director
Resources and Technology Division
Economic Research Service, USDA, Washington

INTRODUCTION

A farm can be regarded as a *food factory* and the criterion for its success is saleable products. Or, it can be regarded as a *place to live,* and the criterion for its success is the harmonious balance between plants, animals and people; between the domestic and the wild; and between utility and beauty - Aldo Leopold.

Sustainable agro-forestry systems refer to production systems that can be maintained over the *long-run* in terms of productivity, profitability, and ecosystem quality. This presentation will focus on the input and output substitution incentives, technologies, and policy alternatives available to producers, policy-makers, and the public. Input substitution in the consumption process, including the willingness-to-pay for environmental quality, increases the demand for sustainable ecosystem quality over time and the substitution of this service for other goods but this component will not be discussed in this presentation.

ROLE OF PRICES AND PUBLIC POLICIES

The main thesis of this presentation is that economic incentives and public polices are the principal driving forces in determining the mixes of outputs, inputs, and technologies developed and adopted. The mixes of outputs, inputs, and technologies adopted have to be compatible with sustaining a viable local and global ecosystem, maintaining a productive agro-forestry sector, and insuring both short- and long-run profitability of local farming enterprises. Also, it is important to note that the current structure of local agriculture may not always be compatible with sustainable agro-forestry systems. Why? Largely because public policies and programmes have not been designed to encourage sustainable systems, but rather, to sustain farmers in farming and promote greater food self-sufficiency.

Typically, these public programmes employ some form of price incentives to encourage commodity production and the expansion of cropland relative to silvicultural production.

Previously forested land is converted to cropland or pastureland at the margin because the prices of commodities increase relative to forest products, i.e., land is shifted to its highest valued use. Additionally, agricultural programmes may contain some measure of supply control. As opposed to output limitations, these measures frequently involve input restrictions, in particular, idling cropland. Land set-aside or retirement programmes have three major negative side-effects. First, cropland becomes more scarce and costly relative to other inputs, encouraging the substitution of chemicals and other inputs for cropland; essentially farming the land more intensively. Second, enhancing the size of the cropland base can become an important long run strategy when some portion of the base is expected to be idled from production. Third, although research programmes typically respond with a lag and are only partially funded by public monies, they tend to focus on enhancing the productivity of the scarce factor, i.e., cropland, with technologies that are land-saving such as chemicals and higher yielding cultivars. Over time further supply control is needed and the environment is subjected to more intensive methods, increased chemical use, and further environmental degradation. Additionally, some agricultural development programmes have encouraged commodity production through land conversion and settlement schemes.

"GETTING PRICES RIGHT"

How does one address the current ecosystem problems and the lack of sustainable agro-forestry systems? The ultimate challenge is "getting prices right" so that more sustainable agro-forestry systems can prosper in localities and regions where forestry is an integral and necessary part of sustainable production systems. It is important to note that "getting prices right" does not necessarily imply a larger role for public policies, but rather, more likely a smaller and different role. First, current commodity policy reform efforts to remove domestic price distortions and to have domestic commodity prices reflect world market price levels is the most critical step.

Second, the external impacts of agricultural and forestry activities on the environment are not necessarily eliminated by policy reform nor will policy reform lead to an instantaneous transition to more sustainable ecosystems. Other public interventions may be necessary because:

-- market prices seldom reflect either the negative or positive environmental impacts of agro-forestry activities; and

-- the transition to more sustainable agro-forestry systems may proceed too slowly without appropriate incentives and assistance.

Public policy options range from subsidies and charges to green incentive payments to regulations and compliance requirements to achieve the desired environmental impacts.

CONSEQUENCES OF PUBLIC PROGRAMMES

Reviewing the historical data on relative prices, input use, and agro-forestry land allocation for the United States, a rather convincing case can be made that producers respond to price incentives over time and that these relative price changes encouraged the substitution of chemical for nonchemical inputs, more chemical-using for less chemical-using crops, and cropland for forest land (USDA).

Why have these changes occurred? First, given the technological alternatives available, producers substitute the relatively cheaper input for the more costly input. Input substitution occurs over time as the cost of one input changes relatively to the cost of another input. For example, as the cost of herbicides for weed control has decreased relative to the cost of labour and machinery used for weed control, producers have substituted herbicides for labour and machinery. Not only have labour costs been increasing over time as a consequence of economic growth and development, but technological advances have significantly reduced the absolute cost and efficacy of chemical weed controls.

Second, the costs of inputs have been changing relative to the price of outputs. As the price of output increases (i.e., frequently related to government policies) relative to the cost of inputs, producers use more variable inputs to increase output (i.e., the derived demand for inputs increases to produce the now more profitable output). For example, as the European Community increased output prices, the intensity of variable input use increased per unit of area and of output. At least in the short-run, producers can only respond by increasing variable inputs, such as chemicals. In the longer run, labour, capital, and to some extent land inputs, can also be adjusted.

Third, as relative output prices change, the more profitable or higher valued crop is substituted for the lesser valued crop or forest product. Because public programmes frequently tend to distort domestic price relationships between commodities and forest products relative to world market prices, farmers have incentives to allocate more land to the production of commodities and less to forestry. This reallocation of land, particularly in on marginal and environmentally sensitive lands, has frequently led to soil erosion and water quality problems, longer term productivity losses, and the demise of sustainable production systems. Thus, to begin to restore more sustainable agro-forestry production systems is going to require continued efforts to remove domestic price distortions.

Although longer term in nature, commodity policy (price) reforms will affect both the magnitude and mix of outputs and thus the intensity of input use as well as the mix of land uses. As price distortions are corrected, forest products should witness relative improvements in profitability in appropriate regions and localities, and the intensity of input use on competing commodities should decrease, both contributing to a more sustainable agro-forestry systems.

EXTERNALITY (NON-MARKET) IMPACTS
OF CURRENT PRODUCTION SYSTEMS

In specific locations and regions, soil erosion adds to food production costs and threatens future food production capacity. Nonpoint source pollution from agricultural lands threatens water supplies, recreation activities, water treatment costs, and wildlife. Land conversion and degradation alter the aesthetic and economic environment of the countryside, global climate patterns, and the general quality of life. Such problems have been caused by a variety of factors: public programmes, lack of clearly defined property rights, unrecognized environmental and wildlife damages, and failure to include environmental and health damages in market prices or costs of commodity products. If we are to establish or re-establish sustainable agro-forestry systems, we have to adopt long term management techniques that recognize the complex ecosystem interrelationships between trees, plants (including crops), animals (including livestock), other resources, and people. Alternatively, this sustainable agro-forestry system has to consider the production of multiple products and the production of the optimal combination of goods and services. Unfortunately, such an optimal combination will not come about without appropriate scientific management practices being pursued by both individuals and institutions.

INCORPORATING EXTERNALITY (NON-MARKET) IMPACTS
INTO SUSTAINABLE AGRO-FORESTRY POLICY

As noted above, the first major step is commodity policy reform, but that such reform only offers a partial solution to the problem. Other public policy options are needed to aid the transition to more sustainable practices and to incorporate non-market impacts into the decision calculus of producers.

First, various forms of educational, technical, and financial assistance may be needed to inform producers of sustainable production system alternatives, to help establish sustainable practices, and to provide incentives for the adoption practices constituting more sustainable systems. Such alternatives may include tree-agronomic crop systems such as allycropping and intercropping, tree-animal systems that provide forage and timber benefits, shelterbelts for stabilizing microenvironments, and forest-specialty crop systems.

Second, longer term land retirement systems that encourage tree planting such as the Conservation Reserve Program (USDA) or long term easements could be used to hasten the re-establishment of agro-forestry systems. Such programmes provide annual payments to producers transferring their lands grass and tree cover crops.

A third option would be *green payments* to producers who adopt sustainable management programs for their farms, including the incorporation of agro-forestry practices and systems where appropriate. Such programme payments could be tied to foregone income, area managed and extent (intensity) of management, or alternatively, incentives to make the adjustments and aid the transition to more sustainable systems.

Compliance requirements could mandate the adoption of sustainable agro-forestry systems in regions and localities as a condition for participating in and receiving benefits from other public programmes. Such an approach is similar to the Conservation Compliance provisions of the 1985 Food Security Act. The effectiveness of such provisions is highly dependent on the attractiveness of other public agricultural programmes that leverage such provisions.

Finally, the focus of public research and education programmes can have a major longer term impact. If research resources are focused on the development of practices and cultivars to support sustainable production systems and research advances are disseminated to producers in a timely fashion, significant advances can be made over time. If the scarcity of environmental resources and the importance of maintaining sustainable ecosystems is fully accounted for in research planning, then technologies that augment these scarce resources and support sustainable systems should evolve (Hayami and Ruttan).

CONCLUSIONS

It was not my intent to go into details and specifics in this presentation. Rather, my objective was to provide an overview of the key sources of problems in developing sustainable agro-forestry systems, identifying short and longer run policy solutions, and linking this framework to implications for the environment and the development of sustainable agro-forestry systems.

The ultimate objective is "get prices right." That can be achieved in part by agricultural policy reform. But that is only part of the solution. The non-market benefits of sustainable agro-forestry ecosystems must be included either in adjusted "market prices" or through some other form of government intervention in the decision calculus of public decision makers and producers. Finally, public intervention of some form is likely needed to reduce the transition costs to and develop appropriate technology for sustainable agro-forestry system.

ANNEX

Figure no.

1.   Relative cost of pesticides to total input cost.
2.   Relative cost of fertilizers to total input costs.
3.   Relative cost of pesticides to fuel costs.
4.   Relative cost of pesticides to labor costs.
5.   Relative cost of pesticides to equipment costs.
6.   Cost of inputs and prices of outputs.

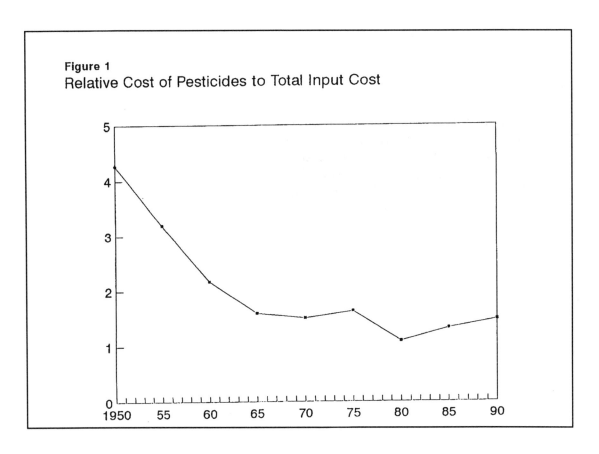

Figure 1
Relative Cost of Pesticides to Total Input Cost

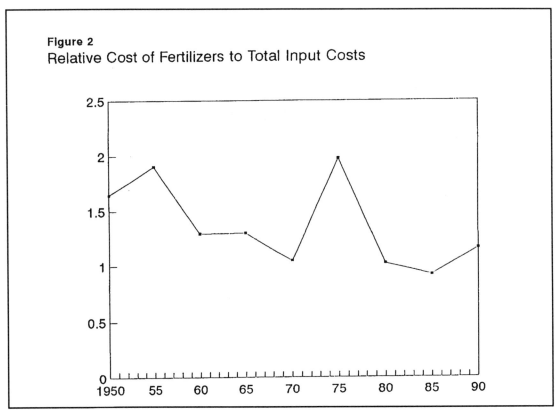

Figure 2
Relative Cost of Fertilizers to Total Input Costs

52

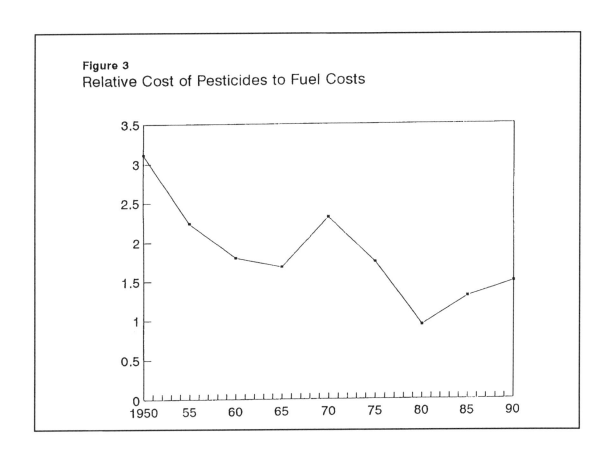

Figure 3
Relative Cost of Pesticides to Fuel Costs

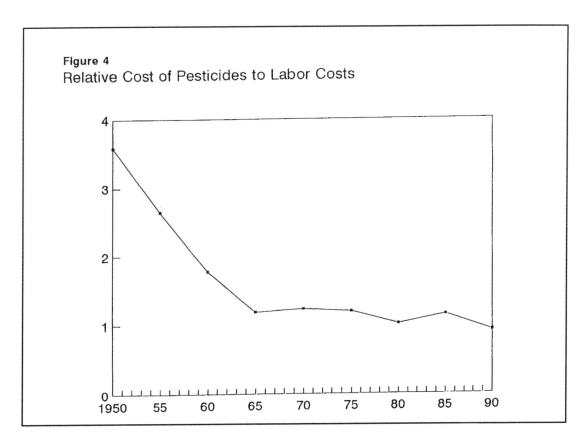

Figure 4
Relative Cost of Pesticides to Labor Costs

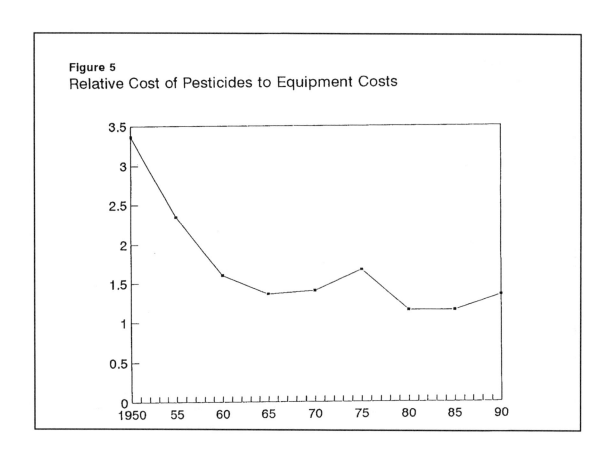

Figure 5
Relative Cost of Pesticides to Equipment Costs

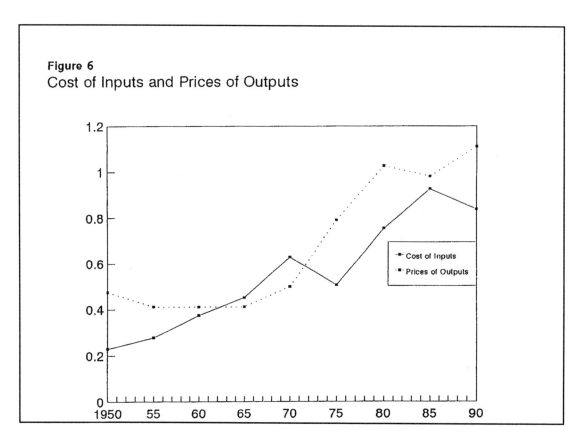

Figure 6
Cost of Inputs and Prices of Outputs

- Cost of Inputs
- Prices of Outputs

AGRICULTURAL POLICY REFORM

AND FARM FORESTRY

John Robert Crabtree, Head
Environmental and Socio-economics Group
Macaulay Land Use Research Institute, Aberdeen

INTRODUCTION

Forestry is the principal alternative to agriculture for the use of land in rural areas. It is typically concentrated in areas of low agricultural productivity, often in upland and mountain regions. The reason for these traditional locations is not difficult to identify. Forestry is characterised as a long-term and illiquid investment which produces relatively low private and social returns to capital, primarily because of its low output value in relation to the time scale of the investment. In the UK, for example, Sitka spruce in state forests produced returns to social capital of under 3 per cent per year from timber output, and this was calculated with land included at zero cost (National Audit Office, 1986). These investment characteristics inevitably mean that forestry, *when evaluated on timber output alone,* is likely to be confined to agricultural land of poor value. Quite large shifts in the relative prices of agricultural and timber products or improvements in forest technology would be required to change this situation. The long-term nature of forestry and the lack of a regular income flow, coupled with quite limited returns to capital have generally made it unattractive for private investment in new planting. Private forestry investment undertaken without subsidy in OECD countries has reflected opportunities for higher returns to capital which typically occur where:

-- higher levels of productivity can be obtained, usually from exotic species with short production cycles;

-- existing forestry has provided capital for subsequent investment;

-- natural regeneration has minimized investment costs; or where

-- other private benefits such as erosion control, flood prevention or the provision of multiple outputs (including sport or amenity benefits) have encouraged landowners to invest.

Against this background, governments wishing to expand their forest base have typically had to intervene either directly through public forest management or indirectly through the provision of incentives to encourage private investors. This has meant that in some countries virtually the entire forestry stock has been created as a result of government intervention. The policy objectives have varied but included such aims as the creation of a strategic reserve, the diversification of rural employment and the maintenance of a supply of timber for domestic processing industries. More fundamentally, it could be argued that intervention was in part directed at overcoming constraints to private investment that were peculiar to the unique characteristics of forestry -- specifically the long time horizon, associated uncertainty and lack of a continuous income flow. In recent years, the focus has shifted to the importance of non-market output from forestry, with intervention in part related to the increased awareness of a role for forestry both in producing environmental outputs and preventing negative environmental externalities from some agricultural land uses (in particular soil erosion and water pollution). There are, however, additional factors which could justify intervention in the forest sector and these relate to situations in which the agricultural sector is supported through price and/or structural polices.

Agricultural commodity support affects the economic returns from forestry and this is particularly relevant on 'marginal' land where commercial forestry is most competitive (Harvey, 1992; Thoroe, 1993). Not only does agricultural commodity support enhance the relative income security of farming as compared to forestry but income benefits to farmers are in part transferred into increases in the prices of inputs, and particularly those in inelastic supply. The main impact of farm support on forestry is through the increase in rents and land prices, although labour and other factor prices could also be affected. Insofar as commodity support systems increase the cost of land for commercial afforestation, returns to forestry investors will be reduced because the higher total cost of establishing a forest (including land purchase) will reduce the Net Present Value produced by forestry. Even where afforestation is undertaken by farmers or landowners themselves, and no sale of land takes place, the potential loss of land value from transferring higher value land out of agriculture can be a significant deterrent to planting since it produces an immediate wealth loss.

A partial justification for some state intervention in forestry might thus be based on compensating for the factor market impacts of agricultural support. This raises an initial question for this workshop - whether policy reform will necessarily reverse impacts of price support on the costs of afforestation. Will reform lead to reductions in input prices for forestry (and particularly in the price of marginal farmland) and result in an increase in the level of new planting? Or will the marginalisation of agricultural land lead to large-scale abandonment, a prospect of concern to policy makers in the EU (Anz, 1993)? While such land use impacts of policy change are clearly important in view of their long-term nature and possible implications for forestry policy itself, the interrelationship between forestry and agricultural policy reform goes much deeper. Intervention to encourage new forestry planting by farmers has been seen as an instrument that can be used in the reform process itself. Most obviously, new planting on agricultural land may assist in the reduction of agricultural output. It may also produce greater positive externalities than some other supply reduction measures (e.g. set-aside) in that it provides farmers with a diversified source of income and offers local employment through the planting and harvesting operations. It could thus have a role in rural development especially in areas of agricultural decline where other employment opportunities may be limited. Such arguments have been frequently rehearsed and, coupled with the potential for environmental gains, have stimulated interest in farm forestry and led the EU and other countries to introduce schemes specifically directed at creating woodlands on farmland.

This paper is concerned with the relationship between agricultural policy reform and forestry. It concentrates mainly on farm forestry and has three specific aims. The first is to assess whether agricultural reform implies increased afforestation of farmland, and if so what form this land transfer out of farming will take. This is analysed in relation to the different possible mechanisms through which reform may take place. A second aim is to take the economic analysis further and ask whether it is desirable to increase the area of afforested land on farms by special measures designed to support new forestry planting. To achieve this requires an analysis of the public benefits from afforestation, including its environmental benefits, its role in rural development and its particular contribution to the adjustment process involved in reform. Finally, the paper examines what instruments are appropriate for delivering farm forestry policy, when policy objectives may vary markedly between countries and zones in relation to the environmental and socio-economic constraints and priorities present.

POLICY OBJECTIVES IN FARM FORESTRY

In addition to a possible role in agricultural policy reform, forestry can produce a diverse range of economic, environmental and social outputs. It has the potential for significant impacts on the environment, and on land conservation, rural employment and the provision of recreational opportunities. Table 1 lists some of the numerous policy objectives that may be used to justify government intervention in farm forestry. Quantifying these benefits is not always possible and converting them to any comparable basis (such as monetary valuations of the benefits) can only be achieved to a very limited extent. Even so, it is helpful to explore the types of effects that farm forestry can produce and their relative importance as between different countries and zones.

Environmental output

Much recent afforestation (particularly with exotic species) has been associated with environmental costs (water quality effects, loss of biodiversity and landscape quality, loss of open space) and a decay in rural structures (e.g. Groome, 1993; Lopez Arias, 1993; Ireland, 1994; Portugal, 1994). There are several facets here, with the environmental benefits (or disbenefits) in part depending on country and location, and on the species and planting arrangements. However, there is no doubt that forestry, with appropriate species, location and management, can have an important role in environmental protection and provide a number of positive environmental externalities. There is a global concern with the loss of forest resources and the opportunity to use the carbon fixing function of growing trees to contribute towards meeting targets for carbon emission stabilisation. But forestry can provide a much wider range of environmental benefits. It can provide benefits to sustainable land management in terms of erosion control and fire prevention. In some OECD countries (e.g. Spain, New Zealand, USA) these impacts are themselves policy priorities with, for example, the recent Spanish announcement of a major programme of afforestation to combat fire hazards and New Zealand's East Coast Forestry Project which aims to afforest 200 000 hectares of eroding and erodible land over a 28 year period. In Northern Europe, benefits to landscape quality, enhanced biodiversity and the provision of additional recreational opportunities dominate the environmental objectives.

Farm scale forestry

One important question is whether farm scale forestry produces a different set of net environmental outputs as compared with large-scale timber-orientated commercial forestry. Farm forestry is likely to be characterised by a mosaic of small discrete blocks, possibly with more species variety than larger-scale planting, although this could depend very much on the structure of farm forestry incentives. In terms of carbon fixing and fire control there would seem to be little difference on a per hectare basis between the effects of farm and large-scale planting. What is primarily important in determining the benefits is the area and location of the planting. With regard to recreational benefits, much would depend on whether planting was associated with rights of public access and this is most likely to occur with state forestry or where special forestry schemes are instituted to provide high levels of recreational benefit. With erosion control the siting of planting is of prime importance and both farm and large-scale planting could be appropriate depending on the situation. Much the same applies to environmental protection, with farm scale forestry having an important role for environmental protection (e.g. shelter and buffer strips) as a coherent part of land use on the farm. But it is in the context of improving landscape and biodiversity that farm forestry may achieve different and potentially more valued benefits than large scale commercial plantations. This mainly reflects the contribution of small-scale, diverse planting in an otherwise agricultural environment.

Farm forestry outputs differ in another significant respect from those of large-scale plantations and this is in their net environmental contribution. As discussed above, commercial timber-orientated forestry is generally only competitive with agriculture on poor quality land. In many situations this land, when un-forested, may produce environmental output in terms of valued wildlife habitats. In such situations the net benefit from afforestation can clearly be negative. However, with farm forestry the scope for planting land which has minimal environmental value or even negative impacts (e.g. water and soil pollution) is much greater.

In these situations the net environmental benefits from farm forestry may potentially be much greater. In the UK, for example, farm forestry has been encouraged partly on the basis of net environmental gains, and a structure of incentives has been used that has encouraged broadleaved planting on arable land.

Rural employment and development

In many situations, afforestation policy has as a prime objective the contribution to rural employment and the maintenance of rural communities. This has been important where land and labour resources were considered to be under-utilised or where agricultural decline and land abandonment have created pressures for alternative rural policies. The particular relevance of farm forestry in this context is that it potentially provides a diversified source of employment for farmers and farm labour, thus assisting in the maintenance of an existing structure of rural employment. While this occurs in situations where there is a farm forestry tradition it is less obvious where this tradition does not exist. Where farmers have unutilised labour they have an incentive to acquire the necessary skills and carry out the operations themselves. Other farmers, however, will either have too great an opportunity cost for their time or prefer to minimise the technical risks by using contractors. The contribution of large-scale, non-farm forestry to rural employment will depend on the source of labour and the opportunities for consistent employment. In some OECD countries (e.g. Ireland, Spain, Italy) the rural development objective in farm forestry is particularly strong, reflecting a wider economic development priority in rural areas -- particularly those characterised by low farm output.

Provision of a supply of timber for processing

Countries developing a domestic timber processing industry may intervene to ensure the provision of a steady stock of timber available for harvest on the basis that a new processing industry cannot realistically be expected to invest in timber production so as to secure its total future supply requirements, given the investment required, and the long time horizon and uncertainties involved. Whatever the merits of this argument it applies mainly to large-scale forestry which is always likely to provide the main source of timber for the processing industry. It is less relevant for farm level investment unless it is in relatively large blocks of suitable species.

Overview

Examining the range of possible objectives in forestry policy emphasises the diversity of situations that exist. It suggests forestry cannot simply be viewed as an instrument of agricultural reform since the outputs and effects of afforestation can themselves make important contributions to other environmental and socio-economic policy objectives. What is quite clear is that countries and zones differ fundamentally in the types of benefits that forestry can provide and this emphasises the importance of setting clear objectives and establishing sets of instruments and conditions that ensure the objectives are achieved. It is to be expected that individual country objectives for forestry policy will differ and that a common policy, covering for example the diversity of farm production and environment in EU will have a number of different objectives in order to address a variety of situations and needs.

FARMER ATTITUDES TO FORESTRY AND UPTAKE OF FORESTRY INCENTIVES

Not a great deal is known or can be inferred either about the general attitudes of farmers to tree planting, or their responses to forestry policy instruments. Farming and forestry are culturally separated in most countries and forestry is a unique investment, radically different from agriculture in its characteristics. With its very long time horizon, lack of constant annual income flow and mix of monetary and non-monetary outputs it is unrealistic to expect that conventional farm management methods based on profit-maximising behaviour will be capable of much precision in predicting behaviour. In addition, research on farmer attitudes and the uptake of forestry incentives has been quite limited, reflecting the novelty of recent policy development to encourage more planting on farms. With large-scale commercial forestry investment, the explanation of investment behaviour is better understood although even there attempts to model planting rates have not generally proved successful (e.g. Whitby, 1983). Despite these difficulties, it is clear that understanding the attitudes of farmers and the constraints that may affect their decisions is central to the achievement of public policy objectives in farm forestry if voluntary instruments are to be used. Farmers' responses both to reform measures and to specific measures designed to stimulate forestry planting will determine the type, location and quantity of planting. They will thus determine both the extent to which farmers themselves benefit from the opportunities offered by woodland planting and the wider environmental and socio-economic impacts of their decisions.

Constraints on planting and response to voluntary measures

A range of structural barriers to farm forestry may exist. Of widespread importance is the fact that farmers may be prevented from planting by land tenure arrangements. This typically occurs under a tenancy agreement either because forestry is not permitted or because the landlord retains the rights to the value of the plantation. From a policy point of view, such institutional constraints not only limit the scope for achieving policy aims but mean that not all farmers may be able to benefit from new measures in forestry. It is well appreciated that farmers frequently lack a forestry tradition and this is important because it may engender negative attitudes to forestry associated with different perceptions of the roles of farming and forestry and the individuals involved in each. It also typically produces a technological barrier to planting because farmers with no forestry tradition lack the necessary expertise and technical confidence to proceed.

There is evidence that farmers without forestry experience are less likely to respond to voluntary incentive payments, and may in some cases have entrenched negative attitudes (Appleton and Crabtree, 1991; Hannan and Commins, 1993). Such factors may be reflected in a preference for the sale of planting land rather than direct engagement in forestry, as occurs for example in Denmark (Denmark, 1994). Traditional concepts of appropriateness in land use may also be a barrier to land use change, in that farmers often regard good quality land, or land clearly suited to farming, as inappropriate for forestry. They therefore show great resistance to planting on land that they feel is better reserved for agriculture. This has been observed almost universally -- for example, in Ireland where there is little forestry tradition (Dhubhain and Gardiner, 1993), and in Finland where there also is a marked reluctance to convert agricultural fields to forest (Selby and Petajisto, 1994).

Financial barriers can also be an important deterrent to farm forestry. Forestry gives no short-term income and it may take 20 years or more before any timber income is produced from new planting. It is therefore unsuited to borrowed finance since, unless special incentive payments are available, there is no income stream to cover interest payments let alone re-payment of capital. This does not mean that indebted farmers will not engage in small-scale forestry without special forms of assistance -- much depends on the additional cash requirements in relation to overall borrowing, income and wealth. The perception of the degree of financial pressure and the income security of the whole business will be an important factor in the decision. However, the absence of an income stream is a very major constraint on the uptake of farm forestry and this has been recognised in the voluntary incentive instrument where both planting grants and annual payments are typically included.

Opportunities for private benefits from tree planting

Unless constraints on planting are intractable, and land tenure arrangements may fall into this category, it is to be expected that farmers may be encouraged to plant by appropriate policy intervention. One clear attraction is the potential for income gains from incentive payments for afforestation. We know little about the methods farmers use to compare the income stream from subsidised forestry with that from agriculture. In particular little is known about the time horizon used for the comparisons or how timber revenue is included in a farmer's assessment of alternative land uses. However, the extent of cost compensation does seem an important element for most farmers -- both in terms of establishment costs and income foregone from farming. In addition to possible income benefits from farm forestry (which are only likely to be substantial in low opportunity cost situations) wealthy farmers may be attracted by any inheritance tax benefits associated with forestry. The extent of any private environmental benefits (environmental protection, landscape, nature conservation and sport) can also be an important element in the decision to plant -- in the UK such benefits have attracted many farmers to take-up voluntary measures, even where this has resulted in a net cost to themselves.

Empirical evidence on attitudes and responses to policy measures

UK Farm Woodland Scheme

Such information as there is on how farmers might react to new policy measures derives either from attitudinal surveys of from evaluation studies on specific farm forestry schemes. One thing is clear --generalisation about farmer behaviour either between or within countries can be misleading given the heterogeneity that can exist both between types of farmers and in the quality of farmland. Studies on UK and Irish farmers illustrate this point. Two studies have evaluated the UK Farm Woodland Scheme (FWS) (Crabtree and Appleton, 1991; Gasson and Hill, 1990). The FWS was a UK scheme launched in 1988 which provided farmers with both planting grants and annual payments -- the aim being to provide compensation both for establishment costs and for the loss of agricultural income. Farmers had a wide range of species choice but broadleaved species received higher planting grants and longer payments to reflect higher differentials in planting costs and longer delays before harvesting.

In view of the payment levels and their structure it was anticipated that much of the planting would be on relatively poor quality lowland (lower compensation was offered to Less-Favoured Area farms) reflecting lower opportunity costs of the withdrawal from agriculture. The assumption was that farmers would base their decisions on the income effects of their entry into the scheme. In fact, this proved not to be the case. Surveys of entrants revealed that most of the planting (73 per cent) was on arable farms (Gasson and Hill, 1990), and the main objectives of farmers in planting were to achieve landscape enhancement and create new wildlife habitats, with timber production much less important. Some farmers were motivated by sport shooting or the provision of shelter. In Scotland (Crabtree and Appleton, 1991), environmental interests again dominated the reasons for uptake with timber even less important than in England. Farmers entering the Scheme were thus primarily influenced by private consumption benefits (environment, amenity, sport) and their impact on farm capital values, and were much less motivated by income benefits from timber. The Scheme particularly appealed to farmers with large profitable arable farms, often those with an additional source of income outside farming and the majority were prepared to suffer a loss in annual income through joining the Scheme. The Scheme produced a widely dispersed mosaic of small woodlands (average of 6.8 hectare per farm) with a predominance of broadleaved planting.

Ireland Forestry Premium Scheme

A voluntary incentive Scheme (the Forestry Premium Scheme) very similar in concept to the UK FWS was introduced into Ireland in 1990 (Government of Ireland, 1991). It provided a mix of planting grants and annual payments with differentials to encourage planting on enclosed land (rather than on poor quality grassland). Previous surveys had shown an absence of any forestry tradition amongst Irish farmers but some commercial forestry potential in view of the large areas of unutilised land of low agricultural productivity, some with potential for high timber yield (more than 22 cubic meters per hectare per year) (Bulfin, 1987; 1993; Dhubhain and Gardiner, 1994). Uptake has been substantial, and farmers have primarily been attracted to plant Sitka spruce on marginal agricultural land, although substantial cultural barriers to afforestation in the west of the country has limited uptake there (Kearney *et al.*, 1993). The driving motivation has been the income gains from the annual incentive payments in relation to minimal opportunity cost, a motivation that has clearly not been heavily constrained by the lack of forestry tradition or expertise. Although timber income should also be achieved it is not evident that this has played a significant part in determining uptake -- presumably because of the time delay and uncertainty involved. Private environmental objectives (landscape, wildlife and amenity improvement) were much less important than in the UK.

Conclusion

The contrast between the responses of farmers in the two countries can hardly have been greater. Since the structure of the schemes was not identical the differences could have been in part dependent on the particular structure of the incentives offered. However, it seems likely that differences between farmers in their commercial and environmental objectives, and differences in the opportunity costs of afforestation, also strongly influenced the uptake characteristics. It suggests that voluntary instruments may need to be very carefully developed in relation to the heterogeneity of farms and farmers if policy objectives are to be achieved. In neither case is there strong evidence that expected timber revenue played a major part in the planting decisions of farmers, presumably because of the associated uncertainty and time delay. It suggests that the level of annual payments in relation to agricultural income foregone is a much more important determinant of uptake rates than the price ratio between farm output and timber. Where short rotation commercial forestry based on exotic species is practised, timber prices and production costs become a more important determinant of planting rates (New Zealand, 1994).

AGRICULTURAL POLICY REFORM

Objectives and instruments

The 1987 OECD Ministerial Communiqué defined the principles and actions for agricultural reform. The long-term objective is to allow market signals to influence, by way of a progressive and concerted reduction of agricultural support, as well as by other appropriate means, the orientation of agricultural production; this will bring about a better allocation of resources which will benefit consumers and the economy in general'.

The ways in which reform will impact on afforestation will depend both on the reform mechanisms used and factors affecting new tree planting. For non-farmer investors, impacts on the price and supply of land will be critical. Whether land prices fell would depend on the nature of the reform process and on expectations about future returns from agriculture and forestry. For farmers, the situation would be more complex. There would be the fundamental impacts of reform on both their income and capital positions to consider, since weakness on either count does not normally provide a sound base for long-term investment in trees. The absence of any income flow for several years after planting means that debts cannot be serviced and that forestry fails to provide a cash income that contributes to living expenses. Much however, depends on the specific opportunities opened up by the reform instruments used and the establishment of measures in support of forestry as part of the adjustment process itself. Where there are farm forestry instruments put in place that contribute to investment costs and provide an annual income it is possible that farmers under financial pressure could turn to afforestation as a way of providing a secure income flow. There is a parallel here with the original five year set-aside scheme in the EC which was partly instituted as a way of assisting arable farmers to cope with the rigours of policy reform. There may be some reluctance to use forestry in this way, because once land is afforested it is normally very costly to convert land back to agriculture and the conditions attached to felling may in any case prevent this. Farmers would thus be engaging in a permanent, rather than temporary shift out of conventional agricultural land use.

It is not the intention to track through all the possible reform instruments in order to demonstrate their effects on these key factors determining new planting. We restrict analysis to a brief discussion of the two main approaches to reform -- those operating through price and revenue, with and without compensation through direct income payments, and those using quantity restrictions to control output.

Price policy reform

Major cuts in institutional support prices (e.g. intervention prices or guaranteed prices under a deficiency payment system) without income compensation will have short-term effects on farm incomes and land prices. The most-indebted farmers will face the greatest adjustment pressures and unprofitable land will cease to be farmed. New Zealand provides the clearest case study of the effects of a radical return to market conditions. Reductions in farm income led to a rapid fall in asset values, reductions in employed farm labour and a rising burden of rural debt (Reynolds and SriRamaratnam, 1990). These effects triggered long-term structural changes in the industry.

External forestry investors

In general, the impact of liberalisation on afforestation as expressed through the level of external (non-farmer) investment depends on whether there is an increased supply of cheaper land, since this should stimulate planting by increasing the return on capital to forestry. This supply of land for afforestation would come partly from insolvency sales and partly from farmers and landowners attempting to raise capital through land sales so as to maintain a viable farming operation on a reduced land area. A marked structural response to the removal of price support and the movement of some land resources out of agricultural use would be anticipated. However, the extent of the land price response to policy reform is difficult to predict. Harvey (1992) calculated that removal of all policies that distort domestic consumption and production would reduce land prices in England and Wales by 46 per cent. Impacts on Scottish hill land (where most UK afforestation has been sited) were less, at 20 per cent. Such reductions in land prices do not increase the commercial returns from forestry all that substantially and certainly not enough to trigger major increases in planting. The increase in land supply may be a more significant factor. There is evidence from Ireland (Bulfin, 1993) that falls in land prices associated with declining agricultural incomes were in part responsible for a surge in afforestation in the late 1980's. In New Zealand, the removal of support to agriculture in the 1980's resulted in a 58 per cent reduction in real farmland values over the six years 1982-1988 (Warren and Sandrey, 1990).

Reductions in land prices and increases in land availability for forestry following price reform would certainly provide conditions for increased commercial investment. Whether this resulted in increased planting would depend on whether other factors such as the level of investment demand or the outlook for wood price were a constraining influence on investment. The fall in land prices in New Zealand did not stimulate an immediate increase in forestry investment. However, it did create a supply of marginal farmland which has facilitated the more recent and substantial increase in planting rates (Roche and Heron, 1993).

Farm forestry investment

With respect to farmer investment in forestry, the effect of major price policy reform is almost certainly negative. A proportion of farmers would leave the industry as part of the adjustment process. For the survivors, lower agricultural incomes and land prices would intensify the search for alternative income sources (including forestry). However, given the investment characteristics of forestry and the opportunities for investment in agricultural property under rapid structural change, it does not follow that farmer forestry planting would increase. Much would depend on policy intervention. Without specific intervention to support farm forestry it would be difficult to identify conditions under which any major planting of trees by farmers would take place. The investment characteristics of forestry would be completely converse to the requirements of farmers under intense price pressure. However, with appropriate incentives, in part perhaps as an adjustment mechanism in reform, there is no doubt that farmers could be encouraged to plant. Annual incentive payments would be an essential pre-requisite since these would provide the attraction for farmers suffering from financial fragility. It must be open to question, however, whether farm forestry is a sensible mechanism for assisting farmers to adjust to price reform *per se*. An equivalent level of annual payment to farmers could almost certainly be provided at lower exchequer cost through the medium of extensification or set-aside instruments.

De-coupled direct income payments

Uncompensated cuts in institutional prices are a rare event. More commonly price policy reform will include compensation to support incomes and the implications for afforestation will then depend critically on how the reform instruments are designed and how they affect incomes and land prices. It is impossible to generalise. With area-based income transfers, land prices would be maintained to the extent that the land was eligible and compensation was complete. Impacts on afforestation would then be minimal. There may be interactions between the instruments used and forestry uptake. For example, direct income payments based on land area may deter farmers from forestry planting if it reduces their qualifying area for payment. Forestry incentives would then need to be pitched at a level that compensated for this loss. On the other hand, the adjustment in production induced by the reductions in commodity price support are likely to lead to some release of land from the industry, even in the context of direct payments, and this would provide opportunities for commercial forestry.

Quantity restrictions

Quotas designed to limit output release resources from the activity under quota - and the resources released will increase over time as technical progress allows the quota output to be achieved with more limited resource inputs. The welfare and resource impacts of quota restrictions have been reviewed in OECD (1990). What happens to the land, labour (and capital) that is released depends critically on the opportunities left for land uses which are outside the quota arrangement or for which additional quota can be purchased or leased i.e. the range of quota controls in operation and whether they operate at the level of the farm or region, and the conditions of transferability etc. A similar situation pertains with land-based quotas, including set-aside. They also remove resources from the farming activity in question but the impact on land use depends on the particular conditions attached to the instrument.

At the farm level there may be situations in which land is intra-marginal because of limited opportunities for use, reflecting a rigid and comprehensive system of quotas on agricultural output. Appropriate forestry incentives could attract this into forestry use. At the industry level there would be pressure for structural adjustment, particularly under a system of tradeable quotas, and this is most likely to provide opportunities for external investment in commercial forestry. One approach which would encourage more farm forestry is to integrate forestry as an option in a set-aside programme. The arguments in favour are that set-aside may itself be a deterrent to the uptake of farm forestry and environmental schemes because of the reluctance of farmers to take more land out of production. Net pubic benefits from the set-aside land are also increased if woodland is planted on some set-aside land. The UK has argued along these lines in the context of the EU set-aside arrangements which currently do not allow forestry (AgraEurope, 1994). However, if set-aside is regarded as a temporary supply control measure, the irreversibility of tree planting and its higher financial costs are also factors that need to be addressed in the decision on set-aside options.

Other measures in reform

Agricultural reform may be accompanied by a range of measures that assist in the reform process and also provide environmental benefits. For example, incentives may be offered to engage in such activities as extensified land use, creation of conservation reserves, and environmental protection and enhancement. It is beyond the scope of this paper to investigate in any detail the possible interrelationships between such measures and farm forestry. It is clear, however, that where a range of options exist for farmers they may not achieve maximum benefit either for the reform process or for the environment unless schemes are spatially targeted to produce greatest complementarity. A range of generally available schemes is likely to result in much less predictable uptake than targeted incentives to achieve pre-determined environmental goals.

There may also be conflicts between reform measures. A good example of this is to be found in the structure of CAP Reform direct payments for the beef sector. Enhanced payments are made to producers stocking at less than 1.4 livestock units per hectare and there is a payment ceiling at 2.0 livestock units per hectare. Producers stocking near these critical levels have a strong incentive to extensify by maintaining or increasing their area of eligible land. This raises the opportunity cost of land for farm forestry and tends to retain grazing land in agricultural production. Such extensification incentives reduce the planting response to any given level of forestry incentive and forestry payments may therefore need to be raised if planting on such farms is desired.

Forestry as a consequence of reform

The impact of policy reform on new planting rates is thus highly dependent on both the economic returns from forestry and the particular instruments used in agricultural reform and their effects on land use opportunities at the farm level. Reform, however managed, does imply land release from farming, and under output restriction measures this would be a progressive release as technological change increased output per unit area. Does this necessarily indicate greater afforestation? One key factor will be whether this land released is through shifts of marginal farmland out of agriculture or whether adjustment takes place through the creation of intra-marginal land within farms which becomes available for alternative uses. In the first case, the extent of forestry planting would depend on the conditions facing external forestry investment, including the extent of regulatory environmental constraints and special forestry incentive measures. It would be surprising if radical reform were not accompanied by shifts of land into forestry, although less-intensive land management systems could well develop as part of the structural adjustment in farming. In the second case of adjustments to land use within farms, there is no guarantee at all that land will be afforested even if unutilised land and labour resources exist. Apart from limited planting for private conservation, environmental and amenity reasons, planting will largely depend on the strength of measures to encourage new planting.

ECONOMIC APPRAISAL OF THE PUBLIC BENEFITS
FROM FARM-BASED AFFORESTATION

Cost-benefit analysis

In some OECD countries public project appraisal relies on cost-benefit analysis to provide the economic basis for certain public investment decisions. This approach has been applied to forestry and it is instructive to examine the extent to which it can inform decision making. As already discussed, forestry can produce a wide range of market and non-market outputs, depending on the type of planting, its location and use. These benefits include timber output, biodiversity gains, recreational use, amenity, environmental protection and socio-economic benefits to rural communities. In some countries or zones the mix may be different (e.g. the importance of game, berry and mushroom production in Sweden) or be focused on one major objective (e.g. fire control in vulnerable Mediterranean areas). If the relevant benefits can be valued in economic terms and then aggregated, we have a measure of the total gross benefit from planting. It then remains to go through the same process on the cost side, including any negative effects (from environmental damage, loss of recreational resources etc.) in order to arrive at a net economic benefit from forestry --usually expressed as a return to capital invested. Although straightforward in principle, there are typically problems in valuing the non-market output and in setting appropriate costs for land and labour inputs. For example, various estimates have been made of benefits that new forestry planting produces through its role in carbon fixing, and a range of estimates of recreational benefits, but there is little information on the value of other environmental benefits (e.g. landscape, biodiversity, erosion and fire control). Even greater difficulties in quantification arise if attempts are made to incorporate the less tangible and more evidently social benefits derived from sustaining local communities, from preventing land abandonment, or from the contribution of forestry to the agricultural reform process itself.

Nevertheless, several such appraisals have been made, and analytical refinement has increased over time as the estimation of environmental and recreational values has been improved. A recent example comes from Pearce (1991) who examined a wide range of different planting types and locations in the UK. Examples of the net economic benefits for five possible systems are given in Table 2. These are calculated with a social value for land fixed at 80% of the market value. The various flow of costs and benefits are discounted at 6 per cent to give present values (£ per hectare), and costs are deducted from the present value of the timber income. Recreational benefits were included by using the consumer surplus estimates of Benson and Willis (1992) which varied from £0 to £424 per hectare depending on location. Most of the accessible forestry in lowland England generated use values of around £200 per hectare. These are, however, gross benefit estimates and do not allow for any use value that may have existed were the land not under forestry. Carbon fixing benefits were admittedly speculative and estimated at £8 per tonne carbon weight.

The following general conclusions can be drawn from the analysis:

-- Economic benefits vary widely, rarely achieving a return to capital exceeding 6 per cent - - the UK government minimum real rate of return for public investment;

-- Carbon fixing benefits are small but significant in the calculation;

-- Benefits tend to be greatest where land is available at low cost (upland sites), and where recreational benefits are highly valued; and

-- Conifers generally outperform broadleaves because of their higher timber revenues.

Only coniferous, mixed conifer/broadleaves and Community (peri-urban) forests can potentially achieve a 6 per cent return on capital, and then only if quite high estimates of the recreational benefits and relatively low land prices are included.

However, this type of analysis can be criticised because it can only include those costs and benefits for which adequate valuation estimates are available. A number of possibly important environmental benefits (and dis-benefits) are not included due to a lack of information, yet it is the perceived environmental gains that have resulted in a re-orientation in policy in many OECD countries towards more environmentally sustainable planting. Certainly, the specific conclusions of Pearce's analysis are somewhat at variance with the shift in UK policy which has been to encourage a greater proportion of broadleaved planting.

Transferred into the context of farm forestry, an equivalent cost-benefit analysis would produce lower returns to capital because recreational benefits would be zero, and timber output may also be lower - a reflection of less experienced management. This focuses attention on the need to achieve other benefits from farm forestry planting if the resource investment is to be justified in economic terms.

Local economic benefits from afforestation

Large-scale planting

In areas of agricultural decline or low productivity, it is often argued that forestry can provide additional opportunities for employment and assist in sustaining fragile communities. Some OECD countries (e.g. Spain, Ireland) have developed forestry programmes with rural development as a prime objective. Typically the emphasis is on timber production and this implies large-scale planting of single species (eucalyptus, Sitka spruce, Pinus radiata), to maximise returns from investment and provide the raw material for added-value wood processing. Such planting has often been associated with environmental damage (and loss of environmental opportunity) but much must depend on what the alternative (or previous) environmental output was from the land. Even so, what is the evidence that rural development benefits are achieved? If forestry displaces no other activity (i.e. takes place on unused or abandoned land) then local gains in employment must occur, and where forestry prevents marginalisation and abandonment there can be social gains through a reduction in the exodus from rural communities. Kearney and O'Connor (1993) calculated the implications for the Irish rural economy of successively larger areas of forestry (additional planting of 10-20per cent of the total land area of Ireland) (Table 3). The predicted net value added and net additional employment are substantial. This reflects two factors that are important pre-requisites for the achievement of major net economic benefits. First, the need for large areas of unutilised land and land of low agricultural productivity such that the impacts on agricultural incomes are predicted to be very small. Second, the need for high growth rates of spruce (18-20 cubic meters per hectare per year). Taken together, these characterise a situation in which returns to social capital (excluding any environmental costs) tend to be high.

Gains in employment are, however, not immediate because by far the most important labour requirement is during harvesting -- which may be 40+ years into the future. With technical progress in timber harvesting, future labour requirements would be reduced. The employment creation associated with forestry may therefore be more limited than appears as first sight but it could still be important in compensating for declining labour use in farming. It is also important to assess the employment creation potential of forestry in relation to other development opportunities in rural areas, since the cost per job created in forestry may still be relatively high (Bateman, 1992; Wibe, 1992). However, in remote, undeveloped or desertified regions forestry may be only obvious alternative to an agriculture in decline and this enhances its role in rural development policy.

Large scale commercial planting can certainly provide a secure source of timber supply for export or processing. But the greatest local benefits are only obtained if locally-sourced labour is used for management and harvesting. In addition, while the existence of commercial planting opportunities can create a demand for planting land and assist in the movement of resources out of agriculture, there may be few benefits for the agricultural population that remains. For example, in Scotland, Mather and Thomson (1993) have shown that when more than 30 per cent of farmland in a region becomes afforested the sense of social isolation for the remaining farmers becomes a factor encouraging a more rapid exodus. Such socially detrimental effects of forestry have also been noted elsewhere (e.g. Dhubhian and Gardiner, 1994). As a policy to assist rural development, forestry may thus have negative as well as positive social impacts, if planting on a substantial scale is envisaged.

Farm Forestry

Where farmers have a forestry tradition and forestry is not displacing agriculture, there are clear employment benefits from expanded farm forestry. However, where no forestry tradition exists, farm forestry may be associated with poor management, low productivity, and excessive harvesting costs, particularly when planting takes the form of widely dispersed blocks. There are certainly employment opportunities for under-utilised farm labour and employed contractors, but much of the long-term rural development benefit depends on the production of saleable crops of timber. The development of arrangements under which timber processors get farmers to engage in new planting on a contract basis could overcome some of the production difficulties if the processors took responsibility for management control and technical advice. But if timber output is in doubt, the principal short-term economic benefits to farmers and farm labour are likely to result solely from the financial transfers that occur in incentive schemes.

FORESTRY AS A MECHANISM IN POLICY REFORM

Intervention to stimulate new forestry planting can have two rather different policy roles. One concentrates on providing suitable conditions for external investors in forestry since this can increase the adjustment of resources out of agriculture. This function of forestry was discussed in the previous section. The second role concerns the involvement of farmers in forestry on farmland. In appraising such a role we need to discuss both the instruments by which farm forestry may be delivered, and the linkage between policy reform and other farm forestry objectives.

Policy instruments

Annual incentive payments

If forestry is to be attractive to a wide range of farmers, annual incentive payments coupled with payments that contribute to planting costs would seem to be essential. Where annual payments do not exist this is known to be a major deterrent to planting since farmers require a consistent source of income. Experience from UK schemes indicates that payments over a period of 10-15 years may be optimal since any extension beyond that is not taken into account in decision making, and pushing payment up-front is a method of increasing cost efficiency and uptake (Crabtree and Appleton, 1992).

Incentive schemes are most cost efficient when the objective is only to encourage planting by a relatively small proportion of the population of farmers. Where much higher uptakes are required, a progressively higher proportion of the payment is transferred as economic rent to the recipient, and this is most marked when land use is heterogeneous. Where high uptakes are required, for example, with schemes to prevent soil erosion or fire hazards, incentive payments may not be effective or may involve high financial cost. In such situations a more regulatory approach in which landowners have an obligation to plant trees has advantages, and this could be coupled with compensation payments to mitigate the costs involved. Spatially targeted schemes are most appropriate for environmental protection since this increases policy effectiveness.

This leads on to another issue for incentive payment schemes which arises with multiple objective forestry (see Table 1). Where multiple objectives exist can incentive schemes be developed that deal satisfactory with the trade-offs involved? For example, is may be that schemes with both environmental and timber production objectives tend to produce compromise mixed species planting that may satisfy neither objective very adequately. It has been observed that an environment managed through intervention may be characterised by less diversity because management tends to focus activities in a narrow desired range (Ehrenberg, 1991). Progressively, what may be required in such situations is a degree of tighter locational targeting with a clearer statement on the policy objectives for specific zones. The development of regulatory strategies which indicate the extent of environmental sensitivity to new forestry planting is a step in this direction (Macmillan, 1993) but it may need to be linked more firmly to incentive mechanisms in order to encourage 'the right trees in the right places'.

Taxation

Special taxation arrangements for forestry are widely used to stimulate investment. They may also be used quite successfully in a converse mode by the imposition of taxation on bare land should this constitute an environmental hazard. Capital and income tax incentives are usually directed at reducing the net-of-tax cost of investment and providing tax-free timber income, sine these can be highly effective investment incentives for external investors. Taxation incentives are less effective for farm forestry.

Technology transfer

Lack of familiarity with the required technology is undoubtedly a barrier to new planting by farmers. Appropriate instruments are grant aid for the employment of consultants to produce plans, and the establishment of free sources of advice and demonstration.

Other instruments

There may be potential for timber processors who wish to increase their security of supply to offer annual payments to growers in exchange for future timber crops. Single species plantations on a suitable scale would be preferred and this will not always fit with other planting objectives. Schemes have also been promoted in which external investors take over a similar role. The objective there is to provide farmers with an annual income and investors with tax-free timber income. Such arrangements are complex and have proved very difficult to mount in practice despite some underlying attractions.

CONCLUSION

The role for *farm forestry* as a mechanism in agricultural reform *per se* would appear to be relatively minor and ancillary to other measures. In order to play a major positive role in adjustment to reform, farm forestry would need to be capable of widespread uptake by farmers on a significant scale. This is unlikely to happen given the barriers to uptake imposed by some types of farm tenure and by the antipathy commonly shown to forestry by a great many farmers. It would also produce major irreversible changes to land use.

Even without evident constraints on uptake, the budgetary cost of farm forestry would be prohibitive if the benefit of farm forestry was restricted solely to its contribution to reform Forestry can only be justified on the basis of its additional environmental and/or development impacts. One universal benefit is carbon fixing. That apart, benefits will be variable, location-specific and not easy to value in economic terms. Where forestry is the main instrument for environmental protection through fire, erosion or pollution control, the case for major forestry programmes is clearly much stronger. Such clear-cut conclusions are difficult to make with respect to the other benefits from forestry and this may partly explain the considerable variation in national target levels for farm forestry under EC Council Regulations 2078/92 and 2080/92. What is clear is that environmental enhancement objectives (landscape, biodiversity, improvements to water quality) usually indicate the need for clear locational targeting to achieve efficiency in policy. The role for farm forestry as an adjustment mechanism on land 'marginalised' or in danger of abandonment as a consequence of reform is highly dependent on the benefits to farmers from annual incentive payments. The economic and social benefits associated with this role for farm forestry need to be appraised in the context of possible alternative large-scale commercial forestry activity on land released from agriculture. Whether farm forestry benefits are largely environmental or produced through rural development, a more targeted approach will progressively be needed if the effectiveness of farm forestry in achieving these objectives is to be maximised.

BIBLIOGRAPHY

AGRAEUROPE (1994), "Woodland Development to Count Against Set-Aside Requirement", Agraeurope July 22nd.

ANZ, C. (1993), "Community Afforestation Policy", in *Afforestation of Agricultural Land*, ed. Volx, K.R. and Weber, N., Commission of the European Communities, Brussels.

APPLETON, Z. E. D. and CRABTREE, J. R. (1991), "The Farm Woodland Scheme in Scotland. An Economic Appraisal", *SAC Economic Report* 27, Aberdeen.

BATEMAN, I. (1992), in *Forests: Market Intervention Failures*, ed. Wibe, S. and Jones, T. Earthscan, London.

BENSON, J. and WILLIS, K. (1992), "The Demand for Forests for Recreation", in *Forestry Expansion: A Study of Technical, Economic and Ecological Factors*, Forestry Commission, Edinburgh.

BULFIN, M. (1987), "Availability of Land for Forestry In Ireland and its Suitability for Sitka Spruce", *Irish Forestry,* 44 (1), 18-31.

BULFIN, M. (1993), "Private Forestry in Ireland: Progress and Problems", in *Afforestation of Agricultural Land,* ed. Volx, K.R. and Weber, N., Commission of the European Communities, Brussels.

CRABTREE, J. R. and APPLETON, Z. E. D.(1991), "Economic Evaluation of the Farm Woodland Scheme in Scotland", *Journal of Agricultural Economics*, 43 (3), 355-367.

DENMARK (1994),"Country Case Study for the OECD Workshop on Forestry, Agriculture and the Environment".

DHUBHAIN and GARDINER (1993), "Small-scale Forestry in the Republic of Ireland", in *Forestry and Rural Development in Industrialised Countries: Where are we going?*, Proceedings of IUFRO Symposium Working Party September 1993, Fredericton, Canada.

DHUBHAIN and GARDINER (1994), "Farm Forestry in the Rural Economy", Agricultural economics Society Conference, Exeter.

EHRENFELD, D. (1991), "The management of diversity : a conservation paradox", in Bormann, F.H. and Kellert, S. R. (1991), (eds) *Ecology, Economics and Ethics.* London : Yale University Press.

GASSON and HILL (1990), "An Economic Evaluation of the Farm Woodland Scheme", F. B. U. Occasional Paper 17, Wye College, Ashford.

GROOME, H. (1993), "Afforestation Policy and Practice in Spain", in *Afforestation, Policies, Planning and Progress*, ed. Mather , A. Belhaven Press, London.

GOVERNMENT of IRELAND (1991), *Forestry Operational Programme*, 1989-1993, Stationary Office, Dublin.

HANNAN, D. F. and COMMINS, P. (1993), "Factors Affecting Land Availability for Afforestation", Economic and Social Research Institute, Dublin.

HARVEY, D. R. (1992), "The Agricultural Demand for Land: Its Availability and Cost for Forestry", in *Forestry Expansion: A Study of Technical, Economic and Ecological Factors*, Forestry Commission, Edinburgh.

IRELAND (1994), Country Case Study for the OECD Workshop on Forestry, Agriculture and the Environment.

JOHNSTON, W. and SANDREY, R (1990), "Land Markets and Rural Debt", in *Farming Without Subsidies, New Zealand's Recent Experience*, ed. Sandrey, R and Reynolds, R. MAF, New Zealand.

KEARNEY, B., HANNAN, D. F., O'CONNOR, R. and COMMINS, P. (1993), *Policy Challenges in Forestry,* Economic and Social Research Institute, Dublin.

KEARNEY, B. and O'CONNOR, R. (1993), *The Impact of Forestry on Rural Communities,* Economic and Social Research Institute, Dublin.

LOPEZ ARIAS, M. (1993), "Experiences of Eucalyptus Afforestation in Spain", in *Afforestation of Agricultural Land*, ed. Volx, K.R. and Weber, N., Commission of the European Communities, Brussels.

MACMILLAN, D. C. (1993), "Indicative Forestry Strategies - An Investment Perspective in the Borders Region of Scotland", *Scottish Forestry,* 47, 83-89.

MATHER, A. S. and THOMSON, K. J. (1993), "The Effects of Forestry Planting on Agricultural Structures", Report to the Scottish Office.

NATIONAL AUDIT OFFICE (1986), *Forestry in Great Britain*, London.

NEW ZEALAND (1994), Country Case Study for the OECD Workshop on Forestry, Agriculture and the Environment.

OECD (1990), *Reforming Agricultural Policies : Quantitative Restrictions on Production, Direct Income Support*, Paris.

PEARCE, D. (1991), "Assessing the Returns to the Economy and to Society from Investments", in *Forestry Expansion: A Study of Technical, Economic and Ecological Factors*, Forestry Commission, Edinburgh.

PORTUGAL (1994), Country Case Study for the OECD Workshop on Forestry, Agriculture and the Environment.

REYNOLDS, R. and SRIRAMARATNAM, S. (1990), "How Farmers Responded", in *Farming Without Subsidies, New Zealand's Recent Experience*, ed. Sandrey, R and Reynolds, R. MAF, New Zealand.

ROCHE, M. M. and LE HERON, R. B. (1993), "New Zealand: Afforestation Policy in Eras of State Regulation and Deregulation", in *Afforestation, Policies, Planning and Progress*, ed. Mather, A. Belhaven Press, London.

SELBY, J. A. and PETAJISTO, L. (1994), *Field Afforestation in Finland in the 1990's*, Finnish Forest Research Institute, Helsinki.

THOROE, C. (1993), "Benefit-Cost Analysis of Agricultural land", in *Afforestation of Agricultural Land*, ed. Volx, K. R. and Weber, N., Commission of the European Communities, Brussels.

WHITBY, M. (1983), "Land Conversion to Forestry in Britain: Some Preliminary Results", Association Européenne des Economistes Agricoles, Séminaire, Nancy.

WIBE, S. (1992), "Policy Failures in Managing Forests", in *Market and Government Failures in Environmental Management*, OECD, Paris.

ANNEX

Table no.

1. Some policy objectives on farm forestry.
2. Cost-benefit appraisals for different forest types (UK).
3. Annual income and employment (over present levels) generated by extra areas of Sitka spruce forest (Ireland).

Table 1. **Some policy objectives on farm forestry**

Environmental protection	Environmental enhancement	Socio-economic and rural development benefits
Fire hazard control Soil erosion control De-intensification of arable production Carbon fixing	Recreation provision Biodiversity increase Landscape improvement	Adjustment to agricultural reform Adjustment to 'marginalisation' Alternative to abandonment Provision of rural development opportunities Diversification of farm activities Provision of a supply of timber for processing

Table 2. **Cost-benefit appraisals for different forest types (UK)**
(Present values (£/ha) at 6% discount rate)

	Spruce Uplands (16m³/ha)	Semi-natural Broadleaves Lowlands	Oak Lowlands (6m³/ha)	Pine Lowlands (14 m³/ha)	Community Forestry
Timber	-458	-3 839	-4 283	-2 605	-3 173
Recreation	268	547	547	476	2 091
Carbon	210	187	246	167	213
Total	**20**	**-3 105**	**-3 490**	**-1 962**	**-869**
Internal rate of return (%)	**6.0**	**0.1**	**1.6**	**3.8**	**4.8**

Source: Pearce (1991).

Table 3. **Annual income and employment (over present levels) generated by extra areas of Sitka spruce forest (Ireland)**

% of land area under forestry including current area of 450 000 ha	Unit 0%	10%	15%	20%
Yield Class (m^3/ha/year)	-	18	20	20
Increase on Present Area	'000 ha	239	583.5	928.0

Additional Annual Net Value Added
 	
    To factory gate
    In processing
	
	£000
	£000
		
		110 173
		58 193
			
			285 853
			151 531
				
				461 533
				244 869

Annual reduction in agricultural income	£000	-	-13 000	-53 000
Annual Total Net Value Added	£000	168 366	424 384	653 402
Net labour requirement	Man years	9 649	24 031	36 368

Source: Kearney and O'Connor (1993).

THE OUTLOOK FOR FARM FORESTRY

AND THE

MARKETS FOR ITS PRODUCTS

UN-ECE/FAO Agriculture and Timber Division, Geneva[1]

INTRODUCTION

This paper is based on two assumptions:

-- that significant areas of agricultural land in OECD countries will become available for alternative uses, mainly as a result of changes in policies supporting food production; and

-- that forestry (afforestation, agro-forestry, etc.) is seen as a potentially important alternative use.

Forests on former agricultural land, like other forests, may be managed with many different objectives, including the provision of industrial wood raw material, of wood for energy, of non-wood products, and of a range of social and environmental services. For farmers, however, the basic objective will be to gain a better or at least more reliable source of income (from the sale of wood or from public subsidy) than they were obtaining from agricultural production or to obtain other benefits, such as improved facilities for hunting and recreation or personal pleasure from a more attractive landscape. The extent to which they can afford to be altruistic and offer social and environmental services to society at large at no charge or without public financial support will vary according to their financial situation, but will in the majority of cases be very limited.

One fundamental difference between agriculture and forestry is the timescale. Farmers expect a return on their sowing or breeding within a matter of months or at most a few years as in the case of horticulture. With some exceptions (e.g. fast-growing plantations) foresters must wait a decade or a generation before the first fruits of their planting begin to appear, and 35 to 100 years or more for the final crop. Of course, once a forest has been established on a sustainable basis, income will flow regularly, although only if the area is sufficiently large, but the establishment period for a sustainable crop is as long as its rotation. Farmers considering afforestation must, therefore, be prepared to make a radical adjustment to their way of thinking about the timing of the return on their investment. Not surprisingly, many are reluctant to take a leap in the dark unless guaranteed in some way of financial help to see them through the years of little or no return on their forest crop.

After a brief review of long-term trends and the outlook in the supply and demand of forest products, the paper will consider the policies relating to the agriculture/forestry interface and conclude with a discussion on the possibilities, difficulties and implications of extending tree cover on agricultural land, as seen from the forestry and forest industry point of view. There are enormous disparities between individual OECD countries with regard to their agricultural and forestry situations, and solutions for the alternative uses of agricultural land will have to be based on local conditions. Nonetheless, with increasing economic interdependence, the need for developing national policies within an international framework becomes correspondingly greater.

THE FOREST SITUATION, WORLDWIDE AND IN THE OECD AREA

The area of forest and other wooded land (FOWL)[2] in the OECD area amounts to approximately 1.2 billion (thousand million) hectares (ha) (UN-ECE/FAO, 1992). This is nearly one quarter of the world total, which is comparable to the OECD's share of total land area and of agricultural land. Its share of world population, however, is appreciably less: below one-fifth (Table 1).

Within the OECD area, North America, with 878 million hectares, accounts for over 70 per cent of the area of FOWL; the European and Pacific countries share the remaining 337 million hectares roughly equally. Of the nearly 4 billion hectares of FOWL in the rest of the world, the countries of Eastern Europe and the former Soviet Union, with 977 million hectares, account for one fourth. Most of the rest is made up of tropical and sub-tropical forests. Globally, FOWL covers almost 40 per cent of the total land area. The percentage is above the world average in North America (43 per cent) and the former Soviet Union (44 per cent) and, within Europe, in the Nordic countries (Finland, Norway, Sweden) with 59 per cent. It is below the world average in the other European and the Pacific countries of the OECD and the rest of the world on average. In terms of FOWL per head of population, the world average of 1.0 hectare *per caput* is exceeded by a large margin in North America, the former Soviet Union and the Nordic countries, while in other European countries it is little more than one quarter of the world average. Because of the huge disparities in population densities, there are exceptions to the generalisation of forest cover being correlated with FOWL *per caput*. The most obvious one is Japan, whose forest cover is amongst the highest in the world at 68 per cent but which has a very low FOWL *per caput* of 0.2. Australia is at the other extreme with 19 per cent forest cover (much of which being "other wooded land") and 8.5 hectares *per caput*. Country detail is given in Table 2.

Of the three main categories of FOWL given in international statistics, *exploitable forest* is the most important in the OECD area, accounting for 479 million hectares or two fifths of the total. *Other wooded land* makes up a similar area, while *unexploitable forest* covers the remaining one-fifth. "Forest" is defined as land with tree crown cover of more than about 20 per cent of the area[3]; and "other wooded land" as land which has some forestry characteristics but is not forest as defined above, and includes open woodland, scrub, brushland, etc. Exploitable forest is forest on which there is no legal, economic or technical restrictions on wood production, including areas where harvesting is not currently taking place; while unexploitable forest falls into two main categories: (1) forest with severe legal restrictions on wood production, e.g. national parks, nature reserves and other protected areas; and (2) forest where physical productivity is too low or harvesting costs too high to warrant wood harvesting on a commercial scale.

From the wood production point of view, exploitable forest is by far the most important category, accounting for well over 90 per cent of total fellings, and consequently forest inventory data on it are the most comprehensive. With transfers occurring from other categories, the area of unexploitable forest is expanding, while that of other wooded land is falling, mainly as the result of conversion of low quality woodland into productive, i.e. exploitable, forest through reforestation. Unfortunately, statistics are inadequate in most countries to allow these trends to be followed accurately.

Canada and the United States between them account for nearly two-thirds of the area of exploitable forest in the OECD area, with the United States alone having an area of 196 million hectares or 41 per cent of the total. The United States is witnessing a marked change in forest land use, with considerable areas shifting from exploitable to unexploitable forest as a result of environmental protection legislation, linked to such issues as the habitat of the northern spotted owl (USDA Forest Service, 1993). But all countries are facing such pressure to a greater or lesser extent.

Table 3 provides country and regional data on the area of exploitable forest in the OECD area and other temperate zone countries, notably the former Soviet Union. It also shows the ratio between exploitable forest and agricultural land, which may be helpful as one indicator of the potential for afforestation on the latter. A high ratio, as in the Nordic countries and Japan, may indicate that the scope for such transfers may be relatively limited. On the other hand, where the ratio is low, as in several European Union countries, notably Ireland, the scope may be much greater. The *per caput* area of exploitable forest, also shown in the table, may also be a helpful indicator. There is some correlation between this indicator and a country's level of self-sufficiency in forest products. If countries' policies should be to reduce their dependency on imports, they may seek to raise the area of exploitable forest towards what should be a self-sufficiency level which, depending on growing conditions, would usually be between 0.25 and 0.35 hectares *per caput*. In the Netherlands and the United Kingdom it is currently 0.02 and 0.04 hectares *per caput* respectively; the average for the European Union is 0.24 hectares *per caput*.

While ownership of agricultural land in market economy countries is almost entirely in private hands, that of forest land is much more diverse, and the pattern varies considerably between countries as a result of long-term historical and social development. The UN-ECE/FAO classification distinguishes four categories of private ownership: by private wood-processing industries; by private corporations or institutions (religious, educational, pension funds, etc.); farm forest: owned by individuals, families or corporations engaged in agriculture as well as forestry; and other: privately owned not elsewhere specified, including the increasing numbers of "absent owners", who live and work away from their forest holdings. Partly because of this movement, some countries cannot separate the last two categories in their statistics. Public forests are in two main categories: those owned by national, state or regional governments or by government owned corporations; and those belonging to municipalities and communes and other publicly owned forests.

An area of 280 million hectares or two-fifths of forest land in the OECD area is privately owned. Table 4 shows the country variations: even in neighbouring countries, the contrasts can be great, as for example between Canada, with 9 per cent private ownership (much of the public forest is in Provincial hands), and the United States with 72 per cent. Within the European Union, the percentage varies between 90 per cent in Portugal (the highest among OECD countries) and 16 per cent in Ireland. The Nordic countries and Austria have high proportions of private ownership: 71 to 84 per cent. At the other end of the scale, Turkey's forests are almost entirely public. The average for Europe is 62 per cent in private ownership, compared with 36 per cent in North America and 38 per cent in the Pacific countries, and 39 per cent as the average for the OECD as a whole.

As stated earlier, the breakdown of private ownership into different categories is not satisfactorily covered by statistics in some countries. Ownership by forest industries is important in the Nordic countries and North America, especially the United States with 31 million hectares, but relatively unimportant elsewhere. Most private forest is owned by farmers or other non-industry individuals or corporations. The relative importance of farmers and others cannot be accurately assessed, but it is probable that the majority of private forest holdings is owned by farmers, even if their share may be declining in line with farming populations and the shift in populations away from the countryside. In some countries, owners not living in or near their forest properties account for a significant proportion of the total. Special efforts may be needed to motivate such owners to manage their forests.

It is estimated on the basis of incomplete information that there are between 12 and 13 million holdings of forest and other wooded land in the farm and other non-industry categories in European countries of the OECD. Available data are shown in Table 5. Average sizes of holding vary considerably: in the three Nordic countries they average 38.5 hectares, while in the other countries of the European Union for which information is available (excluding Austria) they are only one-tenth of that size (3.8 hectares). In France for example, there are 3.68 million holdings with an average size of 2.9 hectares, and in Spain 4.82 million with an average of 3.7 hectares. These figures are important when considering the potential of intensifying management and increasing yield. Unless some form of cooperation exists, it is difficult for small forest owners to manage holdings on an efficient basis to provide revenue on a regular basis. Even in the Nordic countries, there are an appreciable number of holdings of less than 5 hectares, as Table 5 shows.

No information has been provided on the number and size of forest holdings in North America, Australia or New Zealand. Japan reports the number of holdings at 234 000 with an average size of 10.5 hectares.

Given the growing interest in trends in land use in general and the area of forest in particular arising from the extensive loss of tropical forest to other uses over the past few decades, it is surprising and somewhat unfortunate that reliable data are not available from all temperate zone countries on changes over time in area of forest and other wooded land. Even with some data missing, however, it is clear that for Europe there was a net gain in the area of FOWL between 1980 and 1990. On the other hand, there was a substantial net loss in the United States and a small one in Japan. In Mexico, the annual net loss of forest land is estimated at 670 thousand hectares. Data were not available from New Zealand but it is more than likely that there was a net gain as there was also in Australia. New Zealand has had a dynamic planting programme, nearly all of it reportedly on agricultural land (afforestation). The European countries of the OECD accounted for an estimated 1.4 million hectares out of a total net gain for the region over the decade of an estimated 2.0 million hectares, which was made up as follows:

Changes in area of FOWL (Forest and Other Wooded Land), 1980 to 1990	Europe	United States	Japan
		(million ha)	
Expansion	3.8	1.9	..
of which:			
Afforestation	2.5	..	..
Natural expansion	1.3	..	..
Shift to other uses	1.8	5.1	..
Net gain (loss)	2.0	(3.2)	(0.05)

The estimated 250 000 hectares a year of afforestation in Europe would be almost entirely transfer of non-forest land, mostly agricultural, to forest (rather than to other wooded land); and most of it to exploitable forest with production a principal, although not necessarily only objective. Some afforestation, however, is primarily intended for other purposes, for example, soil protection is a major goal for much of the planting in Spain. Natural extension would, at least in the first phase, result in scrub or open woodland, i.e. other wooded land, but some would eventually develop into productive forest.

FAO data indicate that the annual loss of forest in tropical countries has been running at a rate of over 15 million hectares, but it is not quite clear whether all this is permanent loss or whether some of it will recover a tree cover, e.g. after shifting cultivation. Commercial logging has played a part in this development, but the principal reasons have been socio-economic, basically land hunger by strongly growing populations, as well as clearance for cash crops for export and in drier areas overcutting for domestic fuelwood, still the principal energy source in most developing countries. Among developed countries, the only one reporting a significant decline in forest area is the United States (320 000 hectares a year between 1980 and 1990) According to the latest assessment (USDA Forest Service, 1993), this decline is expected to continue, though at a much slower rate, in the coming decades, primarily to make way for urban development and infrastructure (roads, etc.).

Throughout the OECD area, with the possible exception of Mexico, the volume of forest growing stock has been expanding, even in those countries, such as the United States, with a declining forest area. This expansion has been the result of increment rising more than the wood drain on the forest through fellings. Tables 6 and 7 show the details for OECD countries, the former Soviet Union and Eastern Europe. Among the features to note are:

-- the considerable variations between countries in growing stock per hectare, which depends principally on growing conditions, but also upon management and silvicultural practices. Growing stock is particularly heavy in some European countries (250 m^3 or more per hectare in Austria, Germany and Switzerland), and much lighter in the boreal and Mediterranean-type zones, and also where the average age of stands is low, as in Ireland and the United Kingdom;

-- net annual increment also varies considerably for many of the same reasons. It should be noted that in natural, undisturbed stands, net increment may be around zero, since natural losses from disease, fire etc. offsets the gross increment of the stands. This is still the case for parts of Canada and the United States, as well as the former Soviet Union. Intensifying management raises the NAI, and where this is combined with good growing conditions, high rates of growth can be achieved, for instance the 21.5 m^3/ha/year shown for New Zealand plantations;

-- fellings on exploitable forest were reported in the 1990 FAO/ECE Forest Resource Assessment to be less than NAI in virtually all temperate zone countries, including the OECD area. As Table 7 shows, the only apparent exceptions among OECD countries were the fellings of broadleaved species in Greece and Turkey, as well as in Albania and Poland. In Europe, fellings were 71 per cent of NAI, and in North America (Mexico excluded) an estimated 80 per cent, despite the negligible NAI in the natural forests of the latter for reasons explained above;

-- the average volume of fellings per hectare is another way of looking at harvesting intensity. It is over 3 m³/hectares on average in Europe and the United States, but between 1 and 1.5m₃/hectares in Canada, Australia, Japan and the former Soviet Union.

Over the past half century, growing stock and NAI have reportedly doubled in Europe, both east and west, but part of that expansion has been the result of improvements in the coverage and accuracy of inventory statistics, which for various reasons were conservatively estimated in the past -- and perhaps sometimes still today. Nevertheless, it can be categorically stated that, despite growth in the volumes of wood removed from the forest over that period, there has been an appreciable expansion in Europe's forest in volume terms, and to a much lesser extent in area. Forecasts by European countries submitted for the forthcoming FAO/ECE ETTS V (fifth of the series of long-term European timber trends and prospects studies) indicate that this expansion is set to continue in Europe over the next few decades, with the possibility of average growing stock per hectare reaching 200 m³ by 2040, compared with 139 around 1990. The question being increasingly raised is whether such an increase is in the long-term interests of the health, vitality and environmental well-being of the region's forests, and if not, what is to be done about it. Raising fellings would be one obvious answer, but begs the question of whether the markets could be found for the increased volumes.

LONG-TERM TRENDS AND OUTLOOK IN THE DEMAND FOR AND SUPPLY OF FOREST PRODUCTS, WORLDWIDE AND IN THE OECD AREA

Consumption of the main forest products rose by 50 per cent between 1970 and 1990, both in the OECD area and the rest of the world (FAO, annual; UN-ECE/FAO, 1993). With much slower growth in population in the OECD area, however, the increase in per capita consumption was appreciably greater there than in the rest of the world on average, resulting in a further divergence in per capita consumption levels. As shown in Table 8, consumption in OECD countries is predominantly in the form of processed products (sawnwood, wood-based panels, paper and paperboard), accounting for about 89 per cent of the total in terms of roundwood equivalent in 1990. Elsewhere, especially in the developing regions, fuelwood continues to predominate, with about 72 per cent of the total.

In both the OECD area and the rest of the world, consumption of wood-based panels and paper and paperboard rose more strongly, and that of sawnwood more slowly, than the average for all products over the two decades. Another general feature of trends since 1970 has been the apparent consistency of the rise in world fuelwood consumption, whereas that of processed products has followed the economic cycles, with peaks in the early and late 1970s and late 1980s, and troughs in the mid-1970s, early 1980s and early 1990s. The cyclical swings have tended to be more marked in North America than other regions. Despite or because of that, consumption in North America showed the largest increase between 1970 and 1990 within the OECD area, 60 per cent compared with around 40 per cent for Europe and the Pacific countries.

Comparing consumption at separated points of time may not give a reliable indication of underlying trends, because of the timing of the economic cycles. Charts 1 and 2 show the year-by-year trends from 1978 to 1992 (provisional) in the OECD area, the former Soviet Union, Eastern Europe and the developing regions. Amongst other trends, they show the downturn in consumption of sawnwood and to a lesser extent of paper and paperboard since 1990 and the peaking of fuelwood

88

consumption in the mid-1980s. Sawnwood consumption has been particularly affected by the weakness of demand from the construction sector in many market economies in recent years, as well as the steep decline in the countries with economies in transition, notably the former Soviet Union. Since 1993 there have been increasing signs of recovery in some of these countries, notably the United States, but consumption in 1994 in the industrialised countries as a whole may not recover to the late 1980s level.

Compared with the 50 per cent rise in consumption over the two decades 1970 to 1990, roundwood removals increased appreciably more slowly in the OECD area -- by 32 per cent (Table 9). It is not possible to create a complete materials flow chart to show the reasons for this difference, but among the principal factors were:

-- a strong reduction in wastage of wood raw material accompanied by re-use of processing by-products, notably of sawmilling residues by the pulp and panels industries;

-- a strong increase in waste paper recovery and recycling for paper and paperboard production;

-- improvements in technology and changes in raw material furnish resulting in higher yields, for example increased production and use of higher yield pulps, such as TMP (thermo-mechanical pulp) in partial replacement of chemical pulp;

-- in the case of the European and Pacific countries of the OECD, increased net imports of forest products.

North America increased its removals considerably over the two decades -- by 238 million m^3 or 51 per cent, which brought its share of the total of the OECD area removals to 67 per cent in 1990 compared with 58 per cent in 1970. This expansion supported the substantial increases in both domestic consumption and exports over the period. Given the environmental constraints imposed on logging in recent years in parts of the region, especially the Pacific North-West and British Columbia, it is very doubtful whether such strong growth can continue in the future, unless large price increases stimulate more intensive management and draw out increased supplies from at present under-utilised reserves. As noted earlier, the present level of fellings in North America is less than annual increment in exploitable forests, but this could change if, as is likely, extensive areas of publicly owned forests are withdrawn from harvesting for environmental protection reasons.

Fellings in other parts of the OECD area are also currently less -- and in many countries substantially less -- than increment, and there should be no physical impediment to raising removals in the future. The undercutting in the past could be attributed partly to the unsatisfactory relationship between production costs and roundwood prices, as determined by the international market for forest products, and partly to conservative attitudes towards cutting by owners. Forecasts by European countries, which have been prepared for use in the forthcoming FAO/ECE ETTS V study, point to some further growth in fellings over the coming decades, but at a rate still well below the level of net annual increment.

Given the much slower growth in consumption of sawnwood since 1970 than of other processed forest products, it is interesting to observe that removals of large-sized logs (sawlogs, etc.) almost held their share of total removals over the period. For the OECD area as a whole, removals of sawlogs and veneer logs accounted for 54 per cent in 1969-71 and 53 per cent in 1989-91. In the European countries of the OECD, the share rose from 42 to 46 per cent, while it fell in North America from 60 to 56 per cent and in the Pacific countries of the OECD from 60 to 55 per cent. The main explanation appears to be that the increase in the recycling of industrial wood residues and waste paper slowed down growth in the removals of pulpwood.

In that connection, statistics are not complete on the production and use of industrial wood residues and waste paper, but from the information available, it appears that compared with the rise in total roundwood removals of 32 per cent in the OECD area between 1969-71 and 1989-91, and of industrial wood other than sawlogs, etc. of 22 per cent, recycling of industrial wood residues increased by an estimated 56 per cent and of waste paper by 143 per cent, thereby substantially increasing their share of total fibre furnish at the expense of virgin (green) wood fibre. Given the importance attached in many countries to raising the recovery rate of waste paper and the avoidance of waste, further progress can be expected in the coming decades in the re-use of waste paper and wood residues, even if the law of diminishing returns will apply as recovery rates rise. The implications for forestry in many countries, however, is that demand for smaller dimension roundwood (pulpwood, etc.) will continue to rise more slowly than that for the products made from it.

International trade is the means by which imbalances between national demand and supply are compensated. The OECD area accounts for the major part of world trade in forest products, with exports of US$ 79.6 billion f.o.b. in 1990 or 81 per cent of the world total, and imports of US$ 89.3 billion c.i.f., also 81 per cent of the total. Broadly speaking, two groups of countries are major net exporters of forest products, North America and, within Europe, the Nordic countries and Austria; and two are net importers, the European Union (excluding Austria, Finland and Sweden), and Japan. The net trade figures in Table 10 conceal, however, the importance of the United States and the European Union as both major exporting and importing areas, imports of the United States being mainly from Canada and exports from the European Union being mainly to other European Union Members.

The volumes of exports as a share of production of the main forest products, and of imports as a share of consumption, are shown in Table 11 for the main trading groups and countries. In the case of exports, the major part of Canada's production of sawnwood and paper and paperboard was exported in 1990, and more than a quarter of its output of wood-based panels and woodpulp. A quarter of the European Union's production of panels and woodpulp was exported and nearly a half of its paper and paperboard output. Roundwood exports' share of production (removals) was generally small in OECD countries, with the exception of New Zealand, with nearly one quarter. That country is in the process of building up its wood-processing capacity. With regard to imports, the European Union imported between a third and a half of its consumption of sawnwood, woodpulp and paper and paperboard in 1990, and Australia imported a third or more of its needs for sawnwood, panels and paper and paperboard. While the quantities of all kinds of forest products imported by the United States were large, only for sawnwood, mainly imported from Canada, was the share of total consumption quite high -- nearly a quarter. The commodity pattern of Japan's imports has been changing rapidly in recent years. Traditionally, it imported mainly roundwood and wood chips and maintained large sawmilling, plywood and pulp and paper industries. By 1990, it was importing between 20 and 30 per cent of its consumption of sawnwood, panels and woodpulp, but still only 4 per cent of its paper and paperboard needs. In contrast to most other OECD countries, Japan is still a large importer of roundwood, with 44 per cent of its consumption coming from imports in 1990, although the predominance of roundwood imports has been declining during the early 1990s.

The increasing globalisation of markets for forest products, and the lowering or removal of tariff and non-tariff barriers as a result of the GATT Uruguay Round and other initiatives, are intensifying international competition. Market distortions in the past have also resulted from measures which have lowered the cost of production and exports, such as subsidies for forest road building or low stumpage fees, or quantitative restrictions on exports or imports. While protectionism is unlikely to be entirely eliminated, comparative economic advantage will become increasingly important in determining the location of production, whether of roundwood or of processed forest products, and hence the pattern and volume of international trade. The implications for decision-making regarding afforestation are considerable. If the objective is to produce industrial wood raw material, calculations on the costs of growing it and the scale of operation will need to take into account the international dimension of the market. Afforestation aimed at reducing import dependency, for example, would only be justifiable if the costs of production in the importing country were competitive.

There are no recently published assessments at the global level of future demand and supply of forest products, although a number of regional studies are currently in preparation. For example, work is in progress on the fifth of the series of FAO/ECE studies of European timber trends and prospects (ETTS V), to be published in 1995/96, which will examine the outlook for consumption, production and trade of forest products in Europe, for recycling and for removals of wood. Economic growth in the market economies of western Europe is expected to be higher than the depressed levels of the early 1990s, and to be around 2.8 per cent per annum into the first decade of the 21st century. Output in the transition countries of Eastern Europe in aggregate is expected to recover its 1989 level by 2000, and to grow steadily thereafter.

Consumption of sawnwood and panels is expected to grow slowly (around 1 per cent per annum) in the market economies, with constant real prices, although a trend to long-term loss of market share has been noted for some countries and products. Consumption of paper and paperboard is expected to rise rather faster (about 2 per cent per annum), but there is considerable uncertainty about the outcome of competition between electronic and paper-based communications media.

The use of wood as a source of energy is expected to rise by 50 per cent between 1990 and 2020, but if governments were to take strong action to promote renewable energy, this increase might well be an underestimate. Recovery and recycling of waste paper is expected to continue to rise strongly, until it begins to reach technical ceilings.

ETTS V national correspondents provided estimates of changes in the forest resource and wood supply of their respective countries. They expect, in aggregate, only a small increase in Europe's area of exploitable (productive) forest by 2040 (about 5 per cent), and that fellings, while increasing, will remain at about 70 per cent of forest growth (net annual increment). This under-utilisation of the resource must lead to a further build-up of growing stock per hectare.

In the base scenario being developed for ETTS V, Europe's net imports of forest products are expected to increase by about 50 per cent between 1990 and 2010, but in the opinion of the study's authors, this additional need can be met, notably by wood from highly competitive plantations in other regions. Significant price rises are not expected, especially in view of the large reserve capacity of the European resource. An exception could be prices for certain higher quality timbers, as their availability from not sustainably managed or natural forest resources diminishes.

Studies carried out in the 1980s were unanimous in forecasting that consumption of forest products, at the global and regional levels, would continue to expand up to the year 2000 (Hummel, 1984; UN-ECE/FAO, 1986; Peck, 1991). They suggested that, as in previous decades the volume of growth in world consumption, in terms of roundwood equivalent, would be roughly equally divided between fuelwood and processed forest products, with the developing countries accounting for most of the increase in the former. With regard to processed products, the OECD countries and the formerly planned economies including the former Soviet Union would account for the greater part of growth in volume terms but the developing countries' consumption would rise at a faster rate. Nonetheless, because of the much higher population growth in the latter, their per capita consumption would increase much less than in developed regions, if at all, with a further widening of the per capita levels. Consumption of wood-based panels and paper and paperboard would expand more strongly than of sawnwood, as in the past.

Those analyses that have looked beyond 2000 foresee a continuation of the expansion of consumption during the early part of the 21st century. In the United States, for example, consumption of sawnwood in 2040 is projected to be 33 per cent above the 1990 level and that of structural panels 69 per cent higher, compared with estimated growth in population of 42 per cent (USDA Forest Service, 1993). Over the same period, roundwood supplies are projected to rise by 37 per cent, more slowly for softwoods, faster for hardwoods.

At the international level, the studies were published before the worldwide recession of the early 1990s, but it should be emphasised that long-term analyses should not be influenced by medium-term cyclical movements. There is consensus among economic forecasting institutions that the present phase of recovery may be followed by a period of growth of world GDP averaging 2.5 per cent per year, which would not be much lower than that achieved between 1980 and 1990. Somewhat slower growth than average in the industrialised regions could be compensated by faster growth in the developing regions, led by the emerging economies of Asia (Peck & Descargues, 1994). Changes in paper and paperboard consumption are closely correlated with those in GDP, while those for sawnwood and panels are more directly linked to the major end-use sectors, construction and its ancillaries such as furniture. With regard to the latter, it is expected that an increasing share of investment in construction will go to repairs and maintenance and to meeting the needs of specific segments of the population, such as the growing proportion of the older age-groups and smaller families. In the absence of a sound scientific basis, it may be tentatively stated that the outlook for consumption of forest products in the OECD area in the coming decades is, first, a recovery during the 1990s from the trough of 1992/93, with moderate expansion thereafter.

Given the under-utilisation of the forest resource in most parts of the OECD area, and the further possibilities to expand recycling of residues and waste paper, roundwood supply should be able to meet the expansion of consumption from existing sources without undue difficulty and without compromising the principles of sustainability. Some stresses may arise, however, for example from the withdrawal of productive forest land from harvesting for environmental reasons. The outcome of this could be a tendency for the premium on prices for better quality wood to increase further: this has already been seen as a result of developments in the Pacific Northwest of North America. There remains considerable scope for substitution by other wood or non-wood products, where price differentials become marked.

With regard to historical price trends, Charts 3 to 7 show developments in current and deflated prices between 1962 and 1990 (FAO, 1992). The price indices are aggregated from available country sources of information on specific assortments, and because of possible changes in specifications over time and differences between countries, the reliability of the trends shown cannot be assured. With this reservation, the charts do show some interesting trends. Generally speaking, since 1980 and in several cases over a much longer period, prices for the main forest products have been reasonably stable (allowing for cyclical variations) or slightly downwards. This was the case both for raw materials (sawlogs, pulpwood) and processed products (sawnwood, panels, pulp).

Comparable information on the production costs of roundwood and processed forest products are difficult to collate at the international level, but it would appear that these have followed, or been followed by, prices, with the result that profitability in the sector has not changed significantly (UN-ECE/FAO, 1993a). Generally speaking, profitability has not been sufficient to attract strong investment either to forestry or some of the processing industries, such as sawmilling, although there have been notable exceptions, for example New Zealand's plantations.

Production of industrial wood raw material has been and is likely to remain the single most important function of forestry in the OECD and other developed countries, and the one providing most of the revenue. It is also the case that information about the wood producing function of the forest and the industries and markets for wood and its products is much more readily available than for the non-wood goods and services. The forest is providing, however, a wide range of the latter, a few of them having market values but most of them not. Society in industrialised countries is giving increasing importance, in absolute terms as well as in relation to the wood production function, to these functions, which is being more and more reflected in forest policies and management practices (UN-ECE/FAO, 1992). The functions may be grouped under the following headings: protection (of soil, agriculture, habitations, infrastructure, etc.); water supply, regulation and quality; hunting and trapping; grazing (range); nature conservation, including biodiversity; recreation and other social benefits from the forest; production of non-wood products (berries, nuts, mushrooms, honey, etc.). To this list should be added the role of the forest as carbon fixer and provider of oxygen. Because of the problem of placing values on many of these functions, their absolute and relative importance cannot be determined, but especially in densely populated areas, it is reckoned to be far greater in aggregate than wood production, and to be gaining in importance.

Many of the non-wood functions can only be provided by well-established forests, and the time-lag between establishing new plantations and when they can provide these functions may be as long as for wood production. In considering afforestation, therefore, special attention should be given to those functions of plantations that could have a relatively quick impact: soil protection, shelter from wind, certain types of hunting, grazing and fodder (e.g. in agro-forestry), berries and nuts.

A further possibility to consider is for new plantations to relieve existing forests of part of their wood-producing role, thereby allowing them to be managed more intensively or even entirely for other purposes.

Information is lacking on differences that may exist between the importance attached to the provision of wood and of the various non-wood goods and services as seen from the respective points of view of different groups of forest owners -- State, other public, industry, farm, other private; and the beneficiaries -- individuals, industries and commerce, special interest groups and the public at large. The main difference between public and private forests may be that the former are under a greater obligation to meet the wishes of society with regard to e.g. nature conservation, recreation. In most but not all the OECD countries private forests, as well as public, are accessible to the public for leisure pursuits, including non-commercial picking of fruits, nuts and mushrooms. Except in the still rare cases of subsidisation to provide such services, however, private owners are generally obliged to concentrate their management on income-generating functions, principally wood but also hunting and special products such as Christmas trees. With regard to wood, it could be assumed that the assortment composition provided by private owners, including farm forests, does not differ greatly from the general pattern depicted earlier, apart from producing a greater proportion of assortments for auto-consumption, including fuelwood and wood used on the farm such as fencing material, stakes, poles and coppice produce. Plantations can provide several of these assortments from thinnings, but these products hardly add up to a mass market.

POLICIES AFFECTING THE LONG-TERM SUPPLY OF WOOD AND NON-WOOD GOODS AND SERVICES OF THE FOREST, NOTABLY THOSE RELATING TO THE FORESTRY-AGRICULTURE-ENVIRONMENT INTERFACE

The time has passed in many parts of the OECD area, certainly in Europe, of forest land being converted to other land uses, whether for agriculture or urban and industrial development. One exception is the United States, where further transfer of forest land to urban development and infrastructure is expected (USDA Forest Service, 1993). In most other countries, policies are much more concerned with maintaining or extending the existing forest cover and managing the forests so that they will provide a wider range of goods and services besides wood production, or at least so that these functions will be more effectively integrated with wood production. In practice, this involves forms of management that aim at multiple use and sustainable forestry, including nature conservation and biodiversity, while seeking to maintain the principal sources of income, notably wood output, and improve the net revenue of forest enterprises.

At the same time that forest policy and management are adapting to internal pressures, they are having to face external policy changes that would potentially have even greater impacts (Peck and Descargues, 1995). Some of these are already happening, including changes in policies relating to agriculture, the environment and social welfare, including the easing of unemployment. Others are either under discussion or may be expected to be important in the future: these include policies for the development of alternatives to fossil fuels and, related to that, improving air quality and avoiding the impact of climate change. Forests and forestry have an important role to play in implementing many of these policies, but it is not yet clear how that role could best be enacted.

The Madrid Workshop is aimed at clarifying some of these questions relating to the role of forestry. With regard to agriculture, the starting point is the assumption that, as a result of changes in levels of support intended to bring food supply more closely into line with demand, output of food in the OECD area as a whole will be stabilised, if not actually reduced, and that the area of land needed will be appreciably less than at present. Attempts have been made to assess how much land would be affected and where, and the estimates vary widely. Apart from the transfer of land to other uses -- or to no particular use -- there is also the option to adopt less intensive forms of agriculture or agriculture in combination with other uses, such as agro-forestry and "temperate taungya". For the purpose of this paper it is assumed that most, if not all, OECD countries will be seeking alternatives for some of their agricultural land and that such land will be scattered extensively among a very large number of private properties. Much but not all of it will be on less fertile sites and marginal producing areas, such as remote or mountainous regions. Pressures other than from the side of agriculture may demand the transfer of some farming land in the vicinity of urban areas to other uses, including forestry for mainly non-wood functions (recreation, landscape, amenity).

Afforestation is seen as one of the more attractive alternative uses of surplus agricultural land. It would help to solve what is essentially an agricultural policy problem. Can it also be seen, however, as consistent with forest policy and with the long-term interests both of society at large and of the forest and forest industry sector in general and the individuals involved, in particular the landowners? To answer this question it is necessary to consider *firstly* what are the underlying objectives of forest policy; *secondly* what purposes afforestation would serve; and *thirdly* who would benefit. Regarding the first, they could be paraphrased as being:

"to provide from the forest resource, as efficiently as possible, the diverse needs of society, material and non-material, while respecting the economic and environmental sustainability of the resource".

As for the second, forestry has the role of, and afforestation would be expected to contribute to, providing the following:

-- raw wood material for industry (sawlogs, veneer logs and pulpwood);

-- wood for uses in unprocessed or semi-processed form;

-- biomass for energy (fuelwood, green chips, etc.);

-- non-wood forest products (cork, resin, berries, nuts, mushrooms, honey, game meat, Christmas trees, fodder and grazing, etc.) ;

-- social services (alleviating rural unemployment, maintaining rural fabric, regional development; recreation, hunting, fishing, other sporting and leisure pursuits, etc.);

-- environmental services (protection from avalanches and rock slides, soil protection from wind erosion and floodwater, nature conservation, biodiversity, air and water cleansing, noise abatement, etc.).

As to who would benefit from afforestation, this is a key question closely linked with who should fund it. Broadly speaking, there are two main categories of beneficiaries:

-- Those expecting to gain a reasonable return on their investment from the sale of products or services. Since in most cases the return will not occur until well into the future, and the investment has been made for the benefit of future generations, calculations of the benefit have to take into account appropriate discount rates. What is "appropriate" is a source of never-ending discussion, especially when the public purse is expected to provide support for the investment by lowering the real discount rate in one way or another;

-- "Society" or a given segment of it, when the products or services are seen to benefit it in one way or another but are not marketable. In some countries owners of existing forests are expected to and generally are willing to cover the costs of providing the public with non-marketable goods and services themselves (e.g. by internalising them in the costs of wood production), but it would be too much to expect that those willing to carry out afforestation would do so without adequate fiscal or monetary assistance from public sources.

The beneficiaries of forestry activities, including afforestation, therefore, are expected to be either the investors or "society". But who or what are the investors? In the case of afforestation or other schemes to make forestry play a more important role in agriculture, there are a number of possibilities: (1) the landowner (farmer) may decide to spend his own money or to borrow for forestry schemes on his own land. Sometimes he may have the necessary forestry expertise to carry out the work himself, but often it will involve the use of consultants and/or contractors. He may operate individually or in association with neighbours to benefit from economies of scale, sharing of equipment and labour and eventually marketing of forest products. Forest owners' associations or cooperatives exist in many countries to assist landowners in various ways;

-- The landowner may sell land to an outside party who plans to afforest or undertake more active management of the existing woodland. The purchaser may be an individual, most likely a neighbour, or a commercial concern with interests in investing in forestry. The latter may include forest industries seeking to assure their long-term sources of raw material or other types of investors, who may have little or no direct interest in forestry itself, but who can benefit from it through fiscal or other means. Pension funds are one such type of investor. The state or local authorities may also be land purchasers for forestry purposes. There are cases, notably in New Zealand and the United Kingdom, where the state acquired land for afforestation and subsequently sold the land and the plantations back to the private sector as a going forestry concern;

-- The landowner may enter into some form of agreement with an outside investor for the use of his land, which remains in his ownership, for forestry purposes. The landowner will receive rent from the investor for the use of the land and possibly a share of the returns from the forest crop, while the investor will receive all or most of the returns. The long-term nature of forestry may make it difficult to find a satisfactory basis for this type of leasing arrangement;

-- Where the public authorities, whether national or regional, consider that afforestation or other forestry activity would be for the benefit or welfare of society, whether immediately or at some future date, they may offer financial assistance in various forms, including favourable taxation arrangements, grants or subsidies covering part or even the whole of the investment costs. Much of the private sector afforestation in OECD countries has only occurred because of the existence of such public support, the justification for which in the past was mainly the creation of a domestic reserve of wood raw material and the improvement of the trade balance in forest products and/or the provision of raw material for local industries, but more recently the emphasis has been shifting more towards social and environmental benefits.

Information does not exist on the relative importance of the different types of investment in forestry, including the respective shares of the private and public sectors. It is probably fair to assume, however, that individual landowners, especially agricultural smallholders, have not contributed significantly to afforestation except where they have received substantial public support. Furthermore, such support has sometimes been given for a certain minimum area, in order to increase the likelihood of making the project economically viable. With increasing public interest in the social and environmental functions of forestry, however, the basis for public support could be changing, including the giving of support even for small afforestation schemes.

The earlier statement on forest policy needs further explanation. The phrase "as efficiently as possible" means that society as a whole should get good value for the money it spends, whether it is paying for the goods and services from the forest directly through the market or indirectly through subsidies. Regarding subsidies, most if not all OECD countries have made commitments to contain or reduce the levels of subsidisation of food production. They have also recognized that such subsidisation in the past has contributed significantly to the problem of over-production of some agricultural commodities and has caused a serious drain on the public purse. It is very much to be hoped that some lessons have been learnt from this experience, and the same problems will not be allowed to recur in forestry, i.e. the extent of any support will be carefully controlled.

The phrase "while respecting the economic and environmental sustainability of the resource" encompasses the concepts drawn up in the UNCED Agenda 21. As applied to forestry, the *economic* component means that the sector, including the forest owners and wood-processing industries and commerce, must operate in an economically viable manner with adequate returns on their capital to provide or attract sufficient investment for the long-term future. For some time this has often not been the case, particularly among small forest owners and small and medium-sized forest industries, including sawmills. The *environmental* component encompasses the principles of ecological sustainability, including nature conservation, biodiversity, soil protection and the protection of the environment. Under this heading may also be included issues related to climate change, such as carbon sequestration.

Raw wood material for industry

To oversimplify, the wood-processing industries require raw material of two broad categories: high quality, suitable for the production of joinery and cabinet-making timber, plywood and veneer sheets; and other qualities, of which the main assortments are lower-quality sawlogs for construction-grade and similar qualities of sawnwood and pulpwood for the production of woodpulp, particle board and fibreboard. An important criterion for the latter is that it should be cheap, available in large quantities and of uniform quality and species. This usually implies a fairly large scale of forestry operation, including the planting of monoculture stands. The question immediately arises, especially in Europe with its small agricultural holdings, whether and how this could be achieved, and whether Governments should encourage the transformation of large areas to this type of forestry. Such a movement would, moreover, be counter to the drift of modern thinking, which is towards more varied and site-specific silviculture. If, on the other hand, Governments encourage farmers to grow high quality wood, do the farmers have or can they acquire the necessary skills to do so, and will they accept the long growing period involved?

Afforestation is able to provide pulpwood 10 to 25 years after planting and sawlogs and veneer logs 25 years after or longer, depending on the growing site and species planted. Exceptions exist where the delay can be reduced by the use of fast-growing species (poplar, eucalyptus, Radiata pine), but the sites where these can be grown are fairly limited and they are usually monocultures. Generally speaking, afforestation is suitable for producing wood fibre in bulk, rather than high quality assortments, and in many parts of the OECD area there has been and is likely to continue to be an over-availability of the former. In many European countries, for example, there is a serious backlog in thinning operations, because of their lack of profitability. Often losses on thinnings have to be accepted on the assumption or hope that thinning the stand will lead to better quality of and prices for the final crop at the end of the rotation.

As noted earlier, the problem of the over-availability of small-sized roundwood has been compounded by the environmental pressures to recycle waste paper. There has also been strong competition in the international marketplace for certain products of pulpwood, for example by Brazilian eucalyptus pulp. An exception is provided by New Zealand, where its large-scale Radiata pine planting programme, a combination of reforestation on natural forest areas and afforestation, has been successful in finding outlets for roundwood around the Pacific Rim, notably in Japan and South Korea.

New Zealand is a case where favourable conditions for the growing of one exotic species (Radiata pine has accounted for some 90 per cent of the country's afforestation programme) have combined with the existence of large and expanding markets. It appears to be the exception that proves the rule. Great care is necessary, before embarking on large-scale plantation programmes intended for the production of industrial wood, to ascertain their comparative costs at the international level and the likely existence of domestic or overseas markets. The increasing globalisation of forest products markets means that, even in wood-deficit regions such as western Europe, the profitability of afforestation intended for the production of industrial wood cannot be taken for granted. A further consideration for the afforestation of agricultural land is that the area of individual planting plots will be small and probably dispersed, adding to the production and transportation costs, unless the farmers are prepared to sell or lease their land for afforestation by the State or other large-scale enterprises, or to set up cooperative ventures.

The arguments for and against largescale afforestation for the production of industrial wood raw material may be summarised as follows:

-- *on the positive side:*

.   there is the likelihood of supply restrictions from some traditional sources, including the Pacific Northwest of the United States, British Columbia and some natural tropical forest areas, mainly of the higher qualities of timber. Supply prospects from the Russian Federation are difficult to assess;

.   the strong growth in the emerging economy countries, notably on the Pacific Rim, may cause an appreciable rise in their forest products imports;

.   increasing globalisation and competitiveness of forest products markets may stimulate demand;

.   the construction sector in developing countries, notably the enormous demand for low-cost housing, could offer huge markets for sawnwood and panels;

-- *on the negative side:*

.   there are considerable opportunities for substitution of forest products by non-wood products, if prices were to rise significantly (this has already been seen recently in the United States for sawnwood and panels as a result of both demand and supply factors, but it is uncertain how long-lived those price increases may be);

.   fast-growing plantations in the tropics and sub-tropics are able to offer supply flexibility in the medium term (10-15 years), if demand in the OECD outstripped supply;

.   in the long term, paper consumption may be threatened by electronic communication ("the paperless office", television, video), campaigns to reduce packaging, etc.

While the economic benefits of afforestation for industrial wood production as the unique objective are doubtful for many parts of the OECD area, the situation may be made more attractive where planting can serve other functions which are financially viable, either from sales of products or services or, where society as a whole benefits rather than the landowner, from having the operation subsidised. Some of these possibilities are discussed below, such as shelterbelts, biomass for energy, nature conservation, carbon sequestration, etc.

Wood for uses in unprocessed or semi-processed form

The statistics of the production and use of these assortments have been unreliable and now are virtually non-existent at the international level as a result of changes in the classification of roundwood assortments in the UN Harmonized System. In 1980 Europe (the OECD and other countries) was estimated to be using about 23 million m^3 of "wood used in the rough", of which about 6.5 million m^3 of pitprops, still the largest category despite a rapidly falling market, with telegraph and building poles and other poles, pilings and posts also quite important. These assortments accounted for less than 5 per cent of total consumption of forest products in Europe in terms of m^3 EQ. So far as can be ascertained, the situation is similar in other parts of the OECD area. Wood used in the rough probably represents a more important share of output from small private forest holdings, notably farm forests, than from large public forests, because of auto-consumption of fencing material, vine and other agricultural stakes and poles, sawnwood converted by the farmer himself for the construction and repair of farm buildings, and so on.

Over the years wood has been gradually replaced in many such uses by purchased non-wood materials that are more convenient and more durable, and farmers have tended to lose the skills required or lack the time to produce their own materials from homegrown timber. The trend could probably be reversed and markets rebuilt if farm forest owners could be persuaded of the merits of using wood material which they already have to hand or could quite easily grow themselves. These markets would include agricultural cooperatives, garden centres and other wholesalers and retailers catering for house owners in both rural and urban areas.

The potential for such markets deserves further study, including the incentives and organisational structure required for development and the possible implications for afforestation aimed partly or specifically at them. In the meantime, such outlets will remain a doubtful proposition for afforestation, except on a limited and local scale.

Biomass for energy

Less than two centuries ago, wood was still the primary fuel for both domestic and industrial use, but with the advent first of coal and later of electricity, oil and gas, it was largely replaced, except in certain rural localities. By the early 1970s, fuelwood use in Europe had fallen to around 50 million m^3 a year or less than 10 per cent of total wood consumption and maybe 1 per cent of total energy consumption. The oil price shocks of the early and late 1970s caused reappraisals to be made of the potential for alternative fuels, including biomass, and there has been a gradual trend to diversify energy sources away from fossil fuels. However, this process has not been helped by the weakness of oil prices during the 1980s and early 1990s, and the earlier enthusiasm for developing alternative sources, for example through research programmes under the aegis of the International Energy Agency (IEA), has waned, at least for the time being. These programmes showed the potential for efficient methods of growing, harvesting and utilising wood for fuel, including energy plantations using fast-growing coppicing species such as willow, green chipping and wood-generating heat and power for community use (housing estates, hospitals, schools, barracks, etc.), amongst other possibilities.

Some countries are continuing to press forward with programmes for increasing the use of wood for energy, notably Sweden and the United States, and are prepared to support them to a greater or lesser extent with public funds. Subsidisation will be unavoidable in most cases so long as the price of oil remains low (in 1994 around US$ 15/barrel). It would probably need to at least double before wood could compete without subsidies. Assessments vary greatly about the likely future trend of oil prices: some foresee a level of over US$ 40/barrel within 50 years; others do not expect availability of fossil fuels to be a problem for a long time to come and for any price increases to be modest.

A distinction needs to be made between the production of bioenergy for smallscale use, for example in households, and largescale, such as for electricity generation, community or industry heat and power, etc. For the former, there is the advantage of farmers being able to supply their own energy needs from a fairly small area of forest, although problems may arise in meeting national or local regulations regarding the level of emissions, for example of particulates, volatile organic compounds, haze, etc.), as has proved to be the case in Germany, Switzerland and the United States. Largescale uses not only require the availability of a sufficient area of suitable quality land for energy plantations, where supply from existing woody biomass sources is not enough, but also the development of markets (generating plants, etc.), as is being done in Sweden, which may require some form of support from Government. Costs of meeting emission regulations have also to be considered.

From a technical point of view the use of agricultural land for intensive biomass production is entirely possible. Only certain growing sites, such as better quality, moist soils, would be suitable, and there would usually be need for regular inputs of fertilisers and possibly for irrigation. The scale of annual output would have to be sufficient to supply the needs of the consumer (generating plant). Apart from the cost, which for the time being is the main inhibiting factor, doubts arise as to the sustainability of such systems. The debate also remains open about the most suitable and economic type of biomass to be produced for energy, wood or a number of possible annual agricultural crops. Apart from the costs, preference would also depend on the growing conditions, harvesting and storage systems, the type of fuel to be produced (solid wood, chips, wood flour for pellets, charcoal, ethanol, gas, etc.), and so on.

The case for wood as a source of energy rests primarily on its sustainable availability under appropriate management and on the fact that it is CO_2-neutral, that is to say it causes no net addition to the concentration of CO_2 in the atmosphere, because the photosynthetic process involves taking up this gas and generating O_2.

The prospects for afforestation for energy production will depend to a great extent on the price development of fossil fuels, which in turn will partly depend on government energy policies. While physical availability of fossil fuels does not appear to be a problem for the foreseeable future, especially of coal, arguments for policies to encourage the diversification of energy sources are firstly, geopolitical: dependency on possibly unstable sources of oil; and secondly, environmental: the likelihood of climate change resulting from the build-up of gases of combustion of fossil fuels in the atmosphere. It is beyond the scope of this paper to assess the chances of governments grasping these nettles sufficiently firmly to make a significant impact on the long-term prospects for wood as a significant source of energy and hence of afforestation for this purpose.

Carbon sequestration

The debate on climate change and global warming is by no means resolved. There is no question, however, that the level of CO_2 in the atmosphere has risen considerably over the past century and is now higher than at any time in the past 160 000 years. There is no precedent, therefore, on what its impact may be on the world's climate. Widespread concern, epitomised by the UNCED Convention on Climate Change, is directed at halting and eventually reversing the rise in the atmospheric levels of greenhouse gases. The main strategy must be to reduce their emissions by the use of alternative non-polluting energy sources, as discussed earlier, combined with a halt to tropical deforestation which accounts for about one fifth of the net emission of CO_2 to the atmosphere.

Until a more permanent solution is found, consideration has been given to temporary palliatives such as carbon sequestration, the most obvious way being to increase the volume of biomass on a long-term basis, which in practice means forest. Estimates of the amount of afforestation required worldwide to make a significant impact on the level of atmospheric CO_2 range in the hundreds of millions of hectares. There is probably sufficient land in various parts of the world for this, including substantial areas in the OECD countries, including land no longer required for food production. The expertise for creating and managing such a huge area of new forests could also no doubt be found. The crucial question is the financing, which would have to come from the public purse, mainly of the countries of the OECD, even if it were to be spent to a large extent in the tropical regions, where growing conditions are often better. Unless afforestation can be continued indefinitely, which is unrealistic, it can only help in fixing carbon for a certain period, say a century or so. Thought therefore needs to be given at an early stage about what happens when such forests reach maturity and their carbon-fixing capability becomes limited. They could be left to serve other functions or they could be cropped on a rotational basis like a normal wood-producing forest. In the latter case, in order to extend the period of having carbon fixed, it would be necessary to find uses for the greatly increased volumes of wood that would eventually come onto the market where the accumulated carbon could continue to be stored for a further long period. Studies have shown, in fact, that the use of wood in construction and furniture, for example, is an efficient and large-volume way of storing carbon.

On the face of it, the need for carbon sequestration could be a powerful argument for afforestation, but it poses a number of very difficult questions, not least that of funding. One solution could be to integrate it into strategies involving afforestation with other objectives in view, with some degree of public support being provided specifically for the carbon storage function. If forms of carbon or CO_2 taxation are extended, as is under consideration in some countries, this could be a source of the necessary funding.

Non-wood forest products

A distinction should be made between products that are specific to particular zones and those that are widespread. Important examples of the former are cork, production of which is concentrated in the southern part of the Iberian peninsular, and reindeer cropping in the far north of Scandinavia. The scope for extending such activities to other parts of the OECD area, including afforestation or agro-forestry, is limited. On the other hand, many of the other non-wood products are produced naturally or by intent in forests throughout the OECD area.

Well-established forests, other than monoculture plantations, have a diversity of flora and fauna, some of which are collected (berries, nuts, mushrooms, honey, animal fodder, flowers and foliage) or hunted (game animals and birds) by man. For the most part, forest owners are not being paid for providing them, the main exception being for hunting which may generate substantial revenue in certain cases through licenses and fees or the sale of meat and trophies. Newly created forests may be intended to provide non-wood products either by improving habitats for game birds and animals or for producing Christmas trees (which may hardly count as a forest crop but nonetheless are an important source of revenue to landowners in many countries). Otherwise afforestation plans in the past have seldom taken the production possibilities of non-wood products into account, mainly because of their low or non-existent revenue-generating potential.

On the other hand, agro-forestry has potential, although it has so far attracted interest in only a few OECD countries, unlike in many parts of the tropics. New Zealand has had good experience with cattle-grazing under widely spaced, brashed Radiata pine (brashing is the trimming of lower branches with the intention of adding value to the final crop by producing knot-free timber). A similar combination occurs in quite a number of countries with poplar plantations. Another possibility is planting of nut-bearing trees or shrubs (chestnut, walnut, hazel, cobnut), which could be combined with grazing provided pesticide spraying was avoided or carefully controlled.

To be worthwhile to the owner, agroforestry should produce revenue from the combined activities that is at least as great as from a single crop. This means, for example, that if fewer trees are grown at wider spacing to allow cattle grazing or crop growing underneath, the reduced revenue from timber must be at least offset by that from cattle raising. Further research is needed to explore the possibilities of agro-forestry under temperate conditions as a means of diversifying production and maintaining revenue per unit area.

Social and environmental services

This paper is primarily intended to provide the basis for discussion on the potential for extending farm forestry for commercial purposes, notably wood production, but as discussed earlier other functions of forests and forestry are increasing in relative and absolute importance. The environmental aspects are being treated in Panel 3 of the Workshop and in a separate consultant paper, and so need only be touched on here, partly to point out the linkages between the different functions. With some exceptions, social and environmental services are benefitting society as a whole or certain segments of it, rather than the forest owner himself, and consequently the way in which any costs involved in providing those services are covered and shared is of particular concern to both the owner and public authorities.

Social services of forestry may be sub-divided into macro and micro categories. The former include those contributing to broad economic and social objectives, such as regional development.

Particularly in more remote and poorer rural areas, where agriculture is declining and populations are drifting away to the cities, national policy may be to introduce activities that will offer employment and income to the remaining population and maintain the social fabric in the towns and villages. Forestry's role may be through the intensification of existing activities, such as silviculture, wood harvesting and local wood processing; or through the development of new activities, including afforestation, the establishment of new industries and artisan workshops; or through the development of "new" uses of the forest, such as tourism and recreation and the related infrastructure such as hotels, restaurants, parks, campsites, visitor centres and roads. This type of rural development requires considerable public support, generally more than can be provided by local authorities from their own limited resources. It involves an integrated approach in which forestry is one of a range of activities. Afforestation's role will be, in the short term, to provide employment and stable income, and in the longer term, to establish a sustainable source of raw material for local industries and to improve the landscape and the possibilities for recreational and sporting activities for both the local population and visitors. With regard to the employment argument, afforestation may not offer additional jobs per unit area compared with agriculture, except in the first few years of plantation establishment, so may only be attractive when no alternative employment is available locally.

At the micro level, afforestation may be especially useful for landowners in creating diversity of habitat (cover) for wildlife and hence in improving hunting opportunities. Linked to this are the potential *environmental services* that afforestation could provide, for example to build up biological diversity, create nature "corridors" or "networks" to assist migration of animals, birds, insects and plants. An increasingly important role of afforestation is for the creation of social forests within city boundaries or on the periphery intended principally for the enjoyment in one way or another of urban populations, as well as having environmental functions such as air cleansing, noise abatement and so on. The urban-rural interface faces pressure from the demand for agricultural and forest land for urban and infrastructure development. While some changes in land use in this direction are unavoidable, integrated planning involving the maintenance or creation of green spaces (greenbelts, urban forests) can minimise the damage to the environment and landscape.

While some commercial use of forests primarily created for social and environmental purposes may be possible, at least to a limited extent, wood production will seldom be seen as a sufficient source of income to cover the costs of providing other services. If the establishment and management of new stands is not carried out by public bodies, such as the State forest service, grants, subsidies or other incentives will generally have to be offered to landowners or other investors. The levels of support will be a compromise between what is affordable from the public purse and the total costs involved. Account will also have to be taken of the need to follow rules of competition and the avoidance of protectionism laid down in international agreements, such as the GATT Uruguay Round. This problem is not new: a long-standing controversy has existed between the United States and Canada as to whether the latter's wood production is subsidised.

CONCLUSIONS

Farm forestry is a workable solution under most conditions to the problem of finding alternative uses of agricultural land to food production. There are seldom technical or physical barriers to extending tree cover. Furthermore, it may be justified, and indeed be greatly needed, on social as well environmental grounds. This paper has been principally concerned with examining the potential of farm forestry from the point of view of whether the boost which it could give to production of wood and other marketable forest products is also a good reason for the commitment of the large amounts of both private and public funds that would be necessary to achieve the underlying objective of replacing food production by another viable use of the land.

No generalisations about this can be made: situations vary from country to country, and even locally. It has been shown that under conditions where a commercial tree crop can be grown at competitive cost, even without government subsidisation, and where markets are assured, afforestation is fully justified economically and will bring a good return to the investor. It is argued here, however, that such conditions may be the exception rather than the rule in many parts of the OECD area. This is because the existing forests are already being under-utilised and that fuller utilisation is problematic because of international competition and the oversupply of competing raw materials used by the wood-processing industries, notably waste paper. Forest products also face severe competition from non-wood products in many of their markets. This has resulted in the trends and levels of prices for wood raw materials and processed products alike that have not generally made investment in the sector very attractive. The outlook offered in the present analysis is that this situation will not change significantly in the coming decades.

All forecasts are uncertain, and the assumptions on which the findings of this paper are based could prove to have been wrong. Developments that may occur that could alter the outlook include:

-- a marked increase in the importance of markets for forest products outside the OECD area, notably in the emerging economies around the Pacific Rim. This might tighten international markets for industrial wood products and result in increased demand for exports from the OECD area to other regions;

-- there could be stronger expansion of the world economy than has been assumed in this analysis, with a consequently larger rise in demand for forest products;

-- there might be more active acceptance than has been assumed here of the environmental benefit of the greater use of wood products as coming from a sustainable source and whose life cycle is environmentally benign; and

-- events might occur which would cause oil prices to rise markedly and the search for alternative fuels to be reactivated. It is noteworthy that all these scenarios that could have major impacts on the forestry sector, including the prospects for afforestation, are external to the sector.

Until there emerge positive signs of a more favourable investment climate in the forest and forest industry sector in the OECD area, this paper advocates taking a hard-headed approach to the prospects for largescale tree planting for the production of industrial wood raw material. Important exceptions can and will be found, where industrial-type afforestation will prove to offer good returns without the need for help from the public purse. In many cases, however, individual landowners, especially if their holdings are small, as well as those who would subsidise them, should be encouraged to look at other possible reasons for extending tree cover on farms in conjunction with industrial wood production. This paper has listed a series of functions which, individually or in combination, could often form a compelling argument in favour of afforestation, on either a small or extensive scale. Most of these functions serve the public good, however, and landowners stand to benefit little or nothing from providing them, unless financially helped to do so.

National policies to promote afforestation or forms of extending tree cover, such as agroforestry, need to identify clearly what kind of benefits would accrue and to whom. Many of the benefits are still difficult to quantify in economic terms; some have still to be proven as being positive, such as carbon sequestration. Consideration should also be given to the possible negative aspects of afforestation, an aspect which could not be addressed in this paper. Among the main issues raised are:

(1)   *The economic and social conditions under which an expansion of farm forestry would be in the national interest and the extent to which these would be compatible with environmental objectives and with the interests of individual landowners;*

(2)   *The importance of choosing policy instruments to encourage afforestation on agricultural land that achieve the desired results and are cost effective;*

(3)   *The need to make farm forestry more attractive to landowners, not only by means of financial support but also in other ways such as research and development, training and extension, market promotion, facilities for cooperation and so on. The extent to which it is worthwhile, economically and socially, to target very small holdings has to be assessed;*

(4)   *The conditions, especially economic, under which forestry in general and energy plantations on farmland in particular could contribute significantly as an alternative energy source to fossil fuels;*

(5)   *The possibility that carbon sequestration may become an important policy objective of afforestation;*

(6)   *The potential role of agroforestry as a possible alternative to a complete change in land use as a result of afforestation;*

(7)   *If afforestation or other alternative uses of agricultural land are not acceptable, the political, economic and social consequences, as well as the environmental ones, of abandoning such land completely and leaving it to naturally regenerate.*

In the space available in this paper, it has not been possible to tackle all of the issues in depth. The answers are often country-specific. Nonetheless, it is hoped that the general discussion in the panel will serve the purpose of providing the basis for debate on the outlook for farm forestry and the directions that policy makers should be taking to enable forestry to contribute effectively to the solution of the problems facing agriculture in the OECD area at the end of the 20th century.

NOTES

1.   With consultant assistance of Mr. T.J. Peck, European Forest Institute, Joensuu, Finland.

2.   "Forest and other wooded land" (FOWL) is defined as land under natural or planted stands of trees, whether productive or not. It includes land from which forest has been cleared but that will be reforested in the foreseeable future; and areas occupied by roads, small cleared tracts and other small open areas within the forest which constitute an integral part of the forest
       (UN-ECE/FAO, 1992).

3.   For full definitions, see UN-ECE/FAO (1992), *The Forest Resources of the Temperate Zones*, ECE/TIM/62, Vol. 1, Appendix II.

ABBREVIATIONS

ETTS V   Fifth in the series of FAO/ECE European Timber Trends and Prospects Studies (due for publication in early 1996)

FAO   Food and Agriculture Organization of the United Nations

FOWL   Forest and other wooded land

m^3EQ Cubic metres, equivalent volume of wood in the rough

* * * *

EASTERN EUROPE   Albania, Bulgaria, Czech Republic, Hungary, Poland, Romania, Slovak Republic, former Yugoslavia.

EFTA   European Free Trade Association (membership as at 1 January 1995: Iceland, Norway, Switzerland)

EU   European Union (membership as at 1 January 1995: Austria, Belgium, Denmark, Finland, France, Germany, Greece, Ireland, Italy, Luxembourg, Netherlands, Portugal, Spain, Sweden, United Kingdom)

FORMER SOVIET UNION   Data for former Soviet Union were collected before it became separated into new independent states: Armenia, Azerbaijan, Belarus, Estonia, Georgia, Kazakhstan, Kyrgyz Republic, Latvia, Lithuania, Republic of Moldova, Tajikistan, Turkmenistan, Russian Federation, Ukraine, Uzbekistan. Roughly 95 per cent of the former Soviet Union forest resources are located in the Russian Federation.

FORMER YUGOSLAVIA	Data for the former Yugoslavia were evaluated before it became separated into successive states.
NORDIC COUNTRIES	Finland, Norway, Sweden
NORTH AMERICA	OECD countries: Canada, Mexico, United States
PACIFIC COUNTRIES	OECD countries: Australia, Japan, New Zealand
UN-ECE (or ECE)	United Nations Economic Commission for Europe

BIBLIOGRAPHY

HUMMEL, F.C., (Ed.) (1984), *Forest Policy -- A Contribution to Resource Development*, Martinus Nijhoff/Dr W. Junk Publishers, The Hague.

FAO (annual), *Yearbook of Forest Products*, FAO, Rome.

FAO (1992), "Forest Products Prices, 1971-1990", *FAO Forestry Paper* No. 104, Rome.

FAO (1995), "Forest Resources Assessment 1990 - Global Synthesis", *FAO Forestry Paper* No. 124, Rome.

PECK, T.J. (1991), "Medium-term Trends and Prospects for the Consumption of Forest Products", in *Proceedings of the 10th World Forestry Congress,* Ministère de l'Agriculture et de la Forêt, Paris.

PECK, T.J. (1991*a*"), *Meeting the Need for Environmental Protection while satisfying the Global Demand for Wood and Other Raw Materials: A European Perspective"*, for Vancouver B.C. Conference of the Forest Products Research Society, Madison.

PECK, T.J. & J. DESCARGUES (1995), *The Policy Context for the Development of the Forest and Forest Products Sector in Europe,* Forstwissenschaftliche Beitrage der Professor Forstpolitik und Forstökonomie 14, Ecole polytechnique fédérale, Zürich.

RICHARDS, E.G. (Ed.) (1987), *Forestry and the Forest Industries: Past and Future*, Martinus Nijhoff Publishers for the United Nations, Dordrecht.

UN-ECE/FAO (1986), *European Timber Trends and Prospects to the Year 2000 and Beyond*, ECE/TIM/30 (2 vols.), United Nations, New York.

UN-ECE/FAO (1992), *The Forest Resources of the Temperate Zones -- The UN-ECE/FAO 1990 Forest Resource Assessment*, ECE/TIM/62 (2 vols.), United Nations, New York.

UN-ECE/FAO (1993), "Forest Products Statistics 1988-1992", *Timber Bulletin* Vol. XLVI No. 2, United Nations, New York.

UN-ECE/FAO (1993*a*), *Profitability, Productivity and Prices in the Forest Industries in the UN/ECE Region 1974-1990*, ECE/TIM/72, United Nations, New York.

UN-ECE/FAO (1994), "Roundwood Price Statistics and Specifications", *Timber Bulletin* Vol. XLV No. 7 (Special issue), New York and Geneva, 1994.

USDA FOREST SERVICE (1993), *RPA Assessment of the Forest and Rangeland Situation in the United States: 1993 Update* (draft), USDA, Washington, D.C.

ANNEX

Table no.

1.   Population and main land use categories in the OECD area, 1990 Assessment
2.   Population and area of forest and other wooded land in OECD countries, the former Soviet Union and East Europe, 1990 Assessment
3.   Area of exploitable forest in the OECD area, the former Soviet Union and East Europe, 1990 Assessment
4.   Private ownership of forest land (exploitable and unexploitable) in the OECD area, the former Soviet Union and East Europe, 1990 Assessment
5.   Number, area and average size of farm and "other" private holdings of forest and other wooded land, 1990 Assessment
6.   Growing stock and net annual increment on exploitable forest in the OECD area, the former Soviet Union and East Europe, 1990 Assessment
7.   Roundwood fellings on exploitable forest in the OECD area, the former Soviet Union and East Europe, 1990 Assessment
8.   Apparent consumption of the main forest products in the OECD area and the rest of the world in 1969-71, 1979-81 and 1989-91 (averages)
9.   Total roundwood removals in the OECD area and the rest of the world in 1969-71, 1979-81 and 1989-91 (averages)
10.   Net trade in forest products in the OECD area and the rest of the world, 1970, 1980 and 1990
11.   Trade in relation to the production and consumption of the main forest products in selected country groups and countries of the OECD area in 1989-91 (average)

SYMBOLS USED IN TABLES

* Unofficial figure or UN-ECE/FAO estimate
- Nil or negligible
.. Not available

NOTE

In the following tables Cyprus and Israel were included in the OECD total for statistical purposes.

Table 1. **Population and main land use categories
in the OECD area, 1990 assessment**

	TOTAL	Share of world (%)
Population (millions)	950	17.8
Land (excluding water) (million hectares)	3 280	25.4
Agricultural land (million hectares)	1 198	25.0
Forest and other wooded land (million hectares)	1 215	23.7

Source: FAO (1995), UN-ECE/FAO (1992).

Table 2. **Population and area of forest and other wooded land (FOWL) in OECD countries, the former Soviet Union and Eastern Europe, 1990 assessment**

Country	Population ('000)	Total FOWL ('000 ha.)	FOWL as share of total land (%)	FOWL hectares per capita
Iceland	250	134	1.5	0.54
Norway	4 240	9 565	31.2	2.26
Switzerland	6 710	1 186	29.8	0.18
EFTA (3)	**11 200**	**10 885**	**25.0**	**0.97**
Austria	7 710	3 877	47.0	0.50
Belgium-Lux.	10 210	707	21.5	0.07
Denmark	5 140	466	11.0	0.09
Finland	4 990	23 373	76.7	4.68
France	56 440	14 155	26.1	0.25
Germany (W &E)	79 880	10 735	30.7	0.13
Greece	10 050	6 032	46.7	0.60
Ireland	3 500	429	6.2	0.12
Italy	57 660	8 550	28.4	0.15
The Netherlands	14 940	334	9.8	0.02
Portugal	10 530	3 102	35.8	0.29
Spain	38 960	25 622	51.3	0.66
Sweden	8 560	28 015	68.6	3.27
United Kingdom	57 410	2 380	9.9	0.04
European Union (15)	**365 980**	**127 777**	**40.9**	**0.35**
Turkey	58 690	20 199	26.2	0.34
Cyprus and Israel	5 360	404	13.7	0.08
Other Western Europe	**64 050**	**20 603**	**25.7**	**0.32**
SUB-TOTAL EUROPE	**441 230**	**159 265**	**36.5**	**0.36**

Table 2 (continued)				
Canada	26 520	453 300	49.2	17.09
Mexico	88 598	124 057	67.6	1.46
United States	249 970	295 989	32.4	1.18
SUB-TOTAL NORTH AMERICA	**365 088**	**873 346**	**43.4**	**2.41**
Australia	17 090	145 613	19.3*	8.52*
Japan	123 540	24 718	67.8	0.20
New Zealand	3 350	7 472	27.9	2.23
SUB-TOTAL PACIFIC	**143 980**	**177 803**	**21.7**	**1.22**
TOTAL OECD	**950 298**	**1 215 414**	**37.1**	**1.28**
Albania	3 250	1 449	52.2	0.45
Bulgaria	9 010	3 683	33.4	0.41
Czech & Slovak Republics	15 660	4 491	35.8	0.29
Hungary	10 550	1 675	18.2	0.16
Poland	38 180	8 672	28.5	0.23
Romania	23 200	6 265	27.3	0.27
Former Yugoslavia	23 810	9 453	37.0	0.40
Eastern Europe	**123 660**	**35 688**	**31.2**	**0.29**
Former Soviet Union	**288 590**	**941 500**	**44.0**	**3.26**
Other countries	**3 985 033**	**2 927 625**	**31.4**	**0.73**
NON-OECD	**4 397 283**	**3 904 813**	**40.4**	**0.89**
WORLD	**5 347 581**	**5 120 227**	**39.6**	**0.96**

Source: FAO (1995), UN-ECE/FAO (1992).

Table 3. Area of exploitable forest (EF)[a] in the OECD area, the former Soviet Union and Eastern Europe, 1990 assessment

Country	Area of EF ('000 ha.)	EF as % of total FOWL	Ratio of EF to total agricultural land (AL) (AL = 100)	Hectares of EF per capita (ha/cap)
Iceland	--	--	--	--
Norway	6 638	69.4	638	1.57
Switzerland	1 093	92.2	58	0.16
EFTA (3)	**7 731**	**71.0**	**253**	**0.69**
Austria	3 330	85.9	95	0.43
Belgium-Lux.	702	99.3	44	0.07
Denmark	499	100.0	16	0.09
Finland	19 511	83.5	644	3.91
France	12 460	88.0	39	0.22
Germany (W&E)	9 862	91.8	50	0.12
Greece	2 289	37.9	58	0.23
Ireland	394	91.8	7	0.11
Italy	4 387	51.3	25	0.08
Netherlands	331	99.1	14	0.02
Portugal	2 346	76.5	59	0.22
Spain	6 506	25.4	32	0.17
Sweden	22 048	78.7	605	2.58
United Kingdom	2 207	92.7	12	0.04
European Union (15)	**86 829**	**68.0**	**65**	**0.24**
Turkey	6 642	32.9	24	0.11
Cyprus and Israel	168	41.6	27	0.03
Other Western Europe	**6 810**	**33.1**	**24**	**0.11**
SUB-TOTAL EUROPE	**101 370**	**63.6**	**60**	**0.23**
Canada	112 077	24.7	35	4.23
Mexico	27 000*	20.9*	109*	0.30*
United States	195 596	66.1	32	0.78
SUB-TOTAL NORTH AMERICA	**334 673**	**38.1**	**35**	**0.92**

Table 3 (continued)				
Australia	17 005	11.7	38	1.00
Japan	23 829	96.4	443	0.19
New Zealand	2 060	27.6	15	0.61
SUB-TOTAL PACIFIC	**42 894**	**24.1**	**67**	**0.30**
TOTAL OECD	**478 937**	**39.4**	**40**	**0.50**
Albania	910	62.8	127	0.28
Bulgaria	3 222	87.5	52	0.36
Czech & Slovak Republics	4 491	100.0	67	0.29
Hungary	1 324	79.0	21	0.13
Poland	8 460	97.8	45	0.22
Romania	5 413	86.4	36	0.23
Former Yugoslavia	7 768	82.2	55	0.33
Eastern Europe	**31 588**	**88.5**	**46**	**0.26**
Former Soviet Union	**414 015**	**44.0**	**68**	**1.43**

Source: FAO (1995), UN-ECE/FAO (1992).
[a] See definitions in text.

Table 4. **Private ownership of forest land (exploitable & unexploitable)**
in the OECD area, the former Soviet Union
and Eastern Europe, 1990 assessment

Country	Private Forest Land ('000 ha.)	Share of total forest land (%)	Owned by forest industries ('000 ha.)	Farm forest ('000 ha.)	Other private[a] ('000 ha.)
Norway	7 305	84.0	346	(6 959........	...........)
Switzerland	364	32.2	39	325	...........)
EFTA	**7 669**	**78.0**	**385**	**(7 284.........**	**-**
Austria	3 174	81.9	..	2 058	1 ,116
Belgium-Lux.	393	55.7	-	(393.............	...........)
Denmark	339	72.7	35	(304.............	...........)
Finland	14 837	73.8	2 203	(12 634..........	...........)
France	9 771*	74.5	227*	(9 544..........	...........)
Germany (W & E)	4 354	41.5	592*	(3 762*.........	.........)
Greece	567	22.6	121	456	-
Ireland	62	15.7	-	(62...............	...........)
Italy	4 052	60.0	..	..	..
Netherlands	175	52.4	37	20	118
Portugal	2 491	90.4	184	2 307	-
Spain	17 399	71.2	5 713	11 686..........	...........)
Sweden	5 138	61.3	350	(4 788..........	...........)
United Kingdom	1 261	57.1	..	..	..
European Union (15)	**64 013**	**66.3**	**..**	**..**	**..**
Turkey	44	0.5	(44..........	....	....
Cyprus and Israel	15	10.6	-	(15...............	....
Other W. Europe	**59**	**0.7**	**..**	**..**	**..**
SUB-TOTAL EUROPE	71 741	62.2	..	..	..
Canada	23 007	9.3	2 966	(20 041........	...............)
Mexico	7 300*	15.0*	..	..	..
United States	150 448	71.8	30 615	42 060	77 773
SUB-TOTAL NORTH AMERICA[b]	**180 755**	**35.8**	**33 581[b]**	**(139 874.......**	**...............)[b]**
Australia	11 445	28.7	..	..	..
Japan	13 964	57.8	..	..	..
New Zealand	1 924	25.7	..	..	..
SUB-TOTAL PACIFIC	**27 333**	**38.2**	**..**	**..**	**..**
TOTAL OECD	**279 829**	**40.4**	**..**	**..**	**..**

Table 4 (continued)					
Czech & Slovak Reps.	482	10.7	-	114	368
Hungary	9	5.4	-	-	9
Poland	1 471	17.0	27	(1 444..........	..........)
Former Yugoslavia	2 626	31.4	-	2 626	-
East Europe	4 647	13.7	27	(4 620..........	..........)
Former Soviet Union	-	-	-	-	-

Sources : FAO (1995), UN-ECE/FAO (1992).

Notes: a) Other than industry-owned and farm forests, e.g. by corporations, non-farming individuals, etc.
   b) Excluding Mexico

Table 5. **Number, area and average size of farm & "other" private holdings of forest & other wooded land in selected countries, 1990 assessment**

Countries	Number of holdings ('000)	Total area ('000 ha.)	Average size (ha.)	No. of holdings < 5 ha. ('000)
Norway	145.1	6 681	46.0	38.3
Switzerland	251.7	374	1.5	..
Austria	227.8	3 211	14.1	153.1
Belgium-Lux.	123.2	393	3.2	112.6
Denmark	35.7	339	9.0	35.0[c]
Finland	426.3	11 749	27.6	148.3
France	3 677.3	10 707	2.9	3 271.0[b]
Germany (W)	441.9	2 868	6.5	342.1
Greece	1.3	875	691.0	0.6[d]
Netherlands	2.8[e]	175	42.8[e]	..
Portugal	373.7	2 626	7.1	..
Spain	4 822.5	17 889	3.7	..
Sweden	248.9	13 155	53.0	49.2*[a]
Turkey	0.1*[c]	..	202*[c]	..
Japan	234.3	2 454	10.5	198.4
Hungary	3.4[c]	..	3.0[c]	..
Poland	1 378.0	1 471	1.1	1 373.0[d]

Source: FAO (1995), UN-ECE/FAO (1992).

Notes: a) < 20 ha
  b) < 4 ha
  c) 1980 assessment
  d) < 10 ha
  e) Holdings of > 5 ha

Table 6. **Growing stock (GS) and net annual increment (NAI) on exploitable forest (EF) in the OECD area, the former Soviet Union and Eastern Europe, 1990 assessment**

Country	Total GS million (million m^{3a})	GS per ha. (m^3/ha.^{a})	Total NAI (million m^{3a})	NAI per ha. (m^3/ha.)	Coniferous in NAI (%)
Iceland	..	..	..	..	..
Norway	571	86	17.7	2.7	80
Switzerland	360	329	5.8	5.3	70
EFTA (3)	**931**	**120**	**23.5**	**3.0**	**77**
Austria	953	286	16.5	6.6	75
Belgium-Lux.	110	157	5.0	7.2	66
Denmark	54	116	3.5	7.5	73
Finland	1 679	86	67.9	3.6	77
France	1 742	140	65.9	5.3	44
Germany (W&E)	2 674	271	57.2*	5.8*	70*
Greece	149	65	3.3	1.5	47
Ireland	30	76	3.3	8.4	98
Italy	743	169	18.0	4.1	38*
Netherlands	52	157	2.4	7.2	60
Portugal	167	71	11.3	4.8	61
Spain	450	69	27.8	4.3	68
Sweden	2 471	112	91.0	4.1	83
United Kingdom	203	92	11.1	5.0	77
European Union (15)	**11 477**	**132**	**384.2**	**4.4**	**67**

Table 6 (continued)					
Turkey	759	114	20.6	3.1	72*
Cyprus and Israel	7	42	0.5	3.0*	..
Other Western Europe	**766**	**112**	**21.1**	**3.1**	**72***
SUB-TOTAL EUROPE	**13 174**	**130**	**428.8**	**4.3**	**69***
Canada	14 855	133	207.5*	1.9*	68*
Mexico	1 026*	38*	..	..	..
United States	23 092	118	760.0*	3.9*	57*
NORTH AMERICA[d]	**38 973**	**116**	**967.5*[d]**	**3.2*[d]**	**59*[d]**
Australia	1 796	106	35.8*	2.1*	45*
Japan	2 861	120	..	..	..
New Zealand	351	170	26.7[b]	21.5[b]	..
SUB-TOTAL PACIFIC	**5 008**	**116**	**..**	**..**	**..**
TOTAL OECD	**57 155**	**119**	**1 396.3* [c,d]**	**2.8[c,d]**	**57*[c,d]**
Albania	73	80	1.0	1.1	26
Bulgaria	404*	126*	10.6*	3.3*	44*
Czech & Slovak Republics	991	221	31.0	6.9	70
Hungary	229	173	8.2	6.2	17
Poland	1 380	163	30.5	3.6	80
Romania	1 202	222	31.6	5.8	35

Table 6 (continued)					
Former Yugoslavia	1 056	136	27.7	3.6	27
Eastern Europe	**5 335**	**169**	**140.6**	**4.5**	**50**
Former Soviet Union	**50 310**	**122**	**699.9**	**1.7**	**56**

Source: FAE (1995), UN-ECE/FAO (1992).

Notes:
a) Overbark measure
b) Plantations only. For natural forests, the NAI would be very much lower.
c) Excluding OECD Pacific
d) Excluding Mexico

Table 7. **Roundwood fellings on exploitable forest in the OECD area, the former Soviet Union and Eastern Europe, 1990 assessment**

COUNTRY	FELLINGS		FELLINGS per ha.	RATIO OF FELLINGS TO NAI[a]		
	Total	Coniferous		Tot	Con[b]	BL[c]
	(million m^{3d})		(m^{3d}/ha.)	(NAI = 100)		
Iceland	..	..	..	..	..	..
Norway	11.8	11.5	1.8	67	82	10
Switzer-land	5.3	3.9	4.8	91	95	81
EFTA (3)	**17.1**	**15.4**	**2.2**	**73**	**85**	**32**
Austria	17.3	14.2	5.2	79	86	55
Belgium-Lux.	3.7	2.4	6.3	74	73	75
Denmark	2.3	1.4	4.9	65	53	98
Finland	55.9	43.7	2.9	80	81	78
France	48.0	19.0	3.9	73	65	79
Germany (W&E)	42.7	33.1	4.3	75*	83*	50*
Greece	3.4	1.0	1.5	102	61	139
Ireland	1.6	1.5	4.0	48	47	..
Italy	8.0	1.7	1.8	44	25	56
Netherlands	1.3	0.8	3.9	54	58	49
Portugal	10.9	6.6	4.6	96	96	97
Spain	15.0	8.1	2.3	54	43	79
Sweden	57.5	46.6	2.6	63	62	71
United Kingdom	8.1	6.7	3.7	73	79	55
EUROPEAN UNION (15)	**275.5**	**186.9**	**3.2**	**70**	**72**	**68**

Table 7 (continued)						
Turkey	17.2	10.1	2.6	83	68	123
Cyprus and Israel	0.6	0.6	3.8*	..	..	..
Other W. Europe	**17.8**	**10.7**	**2.6**	**84***	**70***	**117***
SUB-TOTAL EUROPE	**310.4**	**212.9**	**3.1**	**71**	**71**	**72**
Canada	151.7	140.0	1.4	73*	99*	18*
United States	619.6	393.5	3.2	81*	91*	69*
SUB-TOTAL NORTH AMERICA[g]	**771.3**	**533.5**	**2.5**	**80***	**93***	**60***
Australia	20.0	7.7	1.2	56	48	62
Japan	35.3*	23.8*	1.5*	..	..	..
New Zealand	14.3	14.1	11.2[e]	52[e]	53[e]	..
SUB-TOTAL PACIFIC	**69.5***	**45.6***	**1.6***	**..**	**..**	**..**
TOTAL OECD[g]	**1 151.3**	**792.0**	**2.5**	**77[f]**	**85[f]**	**64[f]**
Albania	1.6	0.2	1.8	163	77	193
Bulgaria	4.8	1.5	1.8	45	33	54
Czech & Slovak Republics	20.2	15.6	4.9	65	72	49
Hungary	6.1	0.5	4.6	74	39	80
Poland	27.3	20.8	3.2	90	85	109
Romania	16.0	5.5	2.9	50	50	51
Former Yugoslavia	22.0	6.0	2.8	79	82	79
Eastern Europe	**97.8**	**50.1**	**3.1**	**70**	**71**	**68**
Former Soviet Union	**517.6**	**347.7**	**1.3**	**74**	**89**	**55**

Source: FAO (1995), UN-ECE/FAO (1992).

Notes:
    a) Net annual increment
    b) Coniferous
    c) Broadleaved
    d) Overbark measure
    e) Plantations only
    f) Sub-total Europe and North America only
    g) Excluding Mexico

Table 8. **Apparent consumption of the main forest products in the OECD area and the rest of the world, in 1969-71, 1979-81 & 1989-91 (averages)**

	1969-71 (av.)	1979-81 (av.)	1989-91 (av.)	1989-91 Percent share of world total	Percent change between 1969-71 and 1989-91
OECD Total					
- Sawnwood (million m³)	234	241	282	58	+ 21
- Wood-based panels (million m³)	58	75	91	73	+ 57
- Paper & paperboard (million m.t.)	103	132	178	75	+ 73
- Fuelwood (million m³)	84	147	157	9	+ 87
TOTAL (million m³ EQ)	949	1151	1420	39	+ 50
Rest of world					
- Sawnwood (million m³)	185	205	207	42	+ 12
- Wood-based panels (million m³)	13	28	34	27	+162
- Paper & paperboard (million m.t.)	22	37	57	25	+159
- Fuelwood (million m³)	1104*	1324*	1639*	91	+ 48
TOTAL (million m³ EQ)	1526*	1859*	2267*	61	+ 49

Source: FAO (annual), UN-ECE/FAO (1993).

Table 9. **Total roundwood removals in the OECD area and the rest of the world in 1969-71, 1979-81 and 1989-91 (averages)**

	1969-71 (million m^3)	1979-81 (million m^3)	1989-91 (million m^3)	Percent change between 1969-71 and 1989-91	Percent of world total in 1989-91
OECD TOTAL	**794.0**	**921.2**	**1 048.1**	**+32**	**30.2**
SUB-TOTAL EUROPE	260.2	265.2	287.0	+10	8.3
EFTA (3)	11.6	11.2	15.2	+31	0.4
European Union (15)	230.7	231.5	256.1	+11	7.4
Other Western Europe	17.9	22.5	15.6	-13	0.4
SUB-TOTAL NORTH AMERICA	461.8	596.4	699.5	+51	20.2
SUB-TOTAL PACIFIC	72.0	59.6	61.6	-14	1.8
REST OF WORLD	**1 821.0**	**1 990.8**	**2 417.7**	**+33**	**69.8**
Former Soviet Union	381.4	356.3	362.8	-5	10.5
Eastern Europe	85.1	83.6	75.9	-11	2.2
Other	1 354.5*	1 550.9*	1 979.0*	+46	57.1
WORLD	**2 615.0***	**2 912.0***	**3 465.8***	**+33**	**100.0**

Source: FAO (annual)

Table 10. **Net trade[a] in forest products in the OECD area and the rest of the world, 1970, 1980 and 1990 (US$ billion, current prices)**

	1970	1980	1990
OECD TOTAL	**-0.7**	**-2.6**	**-0.7**
Sub-Total Europe	-1.3	-4.5	-5.1
EFTA (3)	-	+0.1	-
European Union (15)	-1.3	-4.4	-4.7
Other	-	-0.1	-0.4
Sub-Total North America[b]	+2.1	+9.7	+14.8
Canada	+2.6	+9.9	+16.0
United States	-0.4	+0.3	-0.8
Sub-Total Pacific	-1.5	-7.9	-10.4
Australia/ New Zealand	-0.1	-0.1	-0.3
Japan	-1.4	-7.8	-10.1
Rest of world	**+0.7**	**+2.6**	**+0.7**
Former Soviet Union	+0.7	+1.9	+2.2
Eastern Europe	-	+0.4	+0.7
Other	-	+0.3	-2.2

Source: FAO (annual).

Notes: a) + = net exports; - = net imports.

b) Mexico included in North American sub-total.

The cif values of imports have been adjusted to fob values in order to make them correspond with fob exports by taking the difference between the world totals of imports and exports (between 10 and 11 per cent).

Table 11. **Trade in relation to the production and consumption of the main forest products in selected country groups and countries of the OECD area in 1989-91 (average)**

A. Exports as % of production					
Country	Roundwood	Sawnwood	Wood-based panels	Woodpulp	Paper and paper-board
OECD TOTAL[a]	**4.4***	**25.7**	**17.5**	**17.4**	**27.2**
Of which:					
EFTA (3)	10.8	18.7	40.3	26.4	69.9
European Union (15)	6.6	32.0	28.6	23.0	48.9
Canada	1.2	69.8	28.9	35.7	67.9
US	3.9	8.7	8.1	9.8	7.8
New Zealand	23.8	1.9	4.7	4.9	4.3
B. Imports as % of consumption					
Country	Roundwood	Sawnwood	Wood-based panels	Woodpulp	Paper and paper-board
OECD TOTAL[a]	**6.3***	**27.8**	**24.9**	**16.6**	**24.5**
Of which:					
EFTA (3)	11.6	25.8	38.8	18.8	56.7
European Union (15)	12.1	43.9	38.5	34.4	47.5
US	0.3	23.3	11.5	8.1	14.7
Australia	..	32.4	37.3	19.5	32.6
Japan	44.0	24.1	30.5	21.3	3.9

Source: FAO (annual), UN-ECE/FAO (database).
Notes: a) Excluding Mexico

LIST OF CHARTS

Chart no.

1.   Apparent consumption of total forest products in industrialised and developing regions, 1978 to 1992

2.   Trends in consumption of the main forest products in industrialised countries, 1978 to 1992

3-10.   Price indices, 1962 to 1990, based on selected national price series, for:

3.   Pulpwood

4.   Chemical pulp

5.   Coniferous logs

6.   Coniferous sawnwood

7.   Non-coniferous logs (temperate zone)

8.   Non-coniferous sawnwood (temperate zone)

9.   Plywood

10.   Particle board

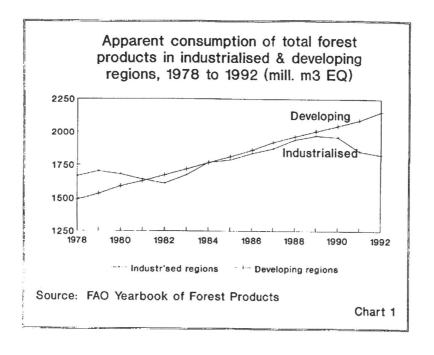

Apparent consumption of total forest products in industrialised & developing regions, 1978 to 1992 (mill. m3 EQ)

Developing

Industrialised

- - - Industr'sed regions   - - Developing regions

Source: FAO Yearbook of Forest Products

Chart 1

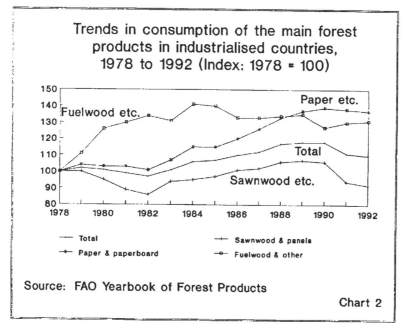

Trends in consumption of the main forest products in industrialised countries, 1978 to 1992 (Index: 1978 = 100)

Paper etc.

Fuelwood etc.

Total

Sawnwood etc.

——— Total   —+— Sawnwood & panels
—*— Paper & paperboard   —□— Fuelwood & other

Source: FAO Yearbook of Forest Products

Chart 2

Chart 3. **Price index for pulpwood, 1962 to 1990,
based on selected national price series
(Index: 1980 = 100)**

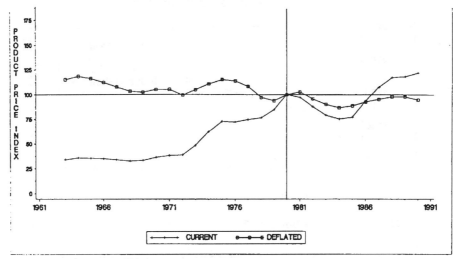

Source: FAO, 1992

Chart 4. **Price index for chemical pulp, 1962 to 1990,
based on selected country price series
(Index: 1980 = 100)**

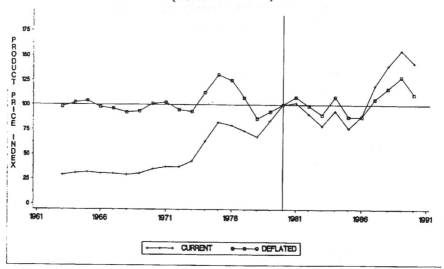

Source: As for Chart 3.

Chart 5.   Price index for coniferous logs, 1962 to 1990,
             based on selected national price series
                    (Index: 1980 = 100)

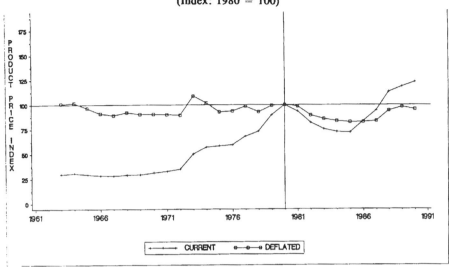

Source: FAO, 1992

Chart 6.   Price index for coniferous sawnwood, 1962 to 1990,
             based on selected country price series
                    (Index: 1980 = 100)

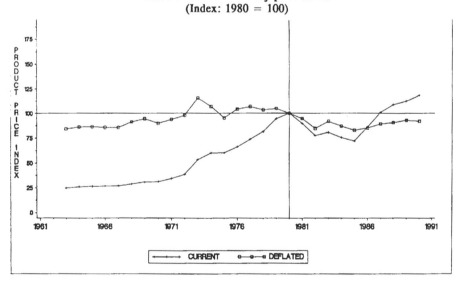

Source: As for Chart 3.

131

Chart 7.   Price index for non-coniferous logs (temperate zone)
            1962 to 1990, based on selected national price series
            (Index: 1980 = 100)

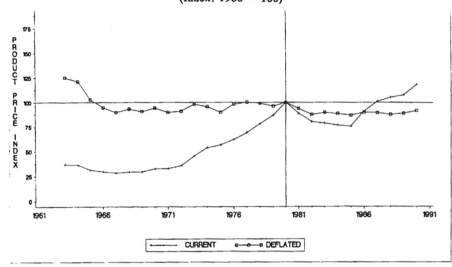

Source: FAO, 1992

Chart 8.   Price index for non-coniferous sawnwood (temperate zone)
            1962 to 1990, based on selected country price series
            (Index: 1980 = 100)

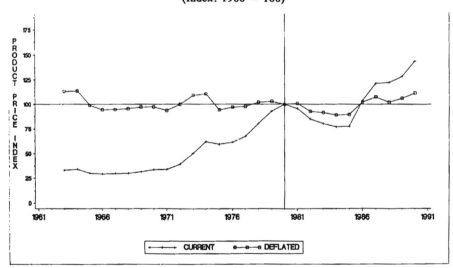

Source: As for Chart 3.

Chart 9.   Price index for plywood, 1962 to 1990,
                based on selected national price series
                    (Index: 1980 = 100)

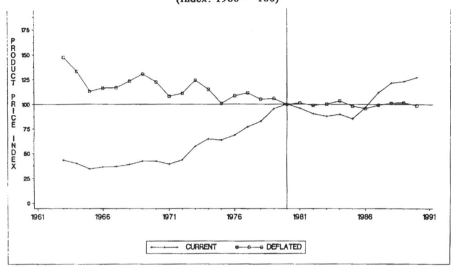

Source: FAO, 1992

Chart 10.   Price index for particle board, 1962 to 1990,
                based on selected country price series
                    (Index: 1980 = 100)

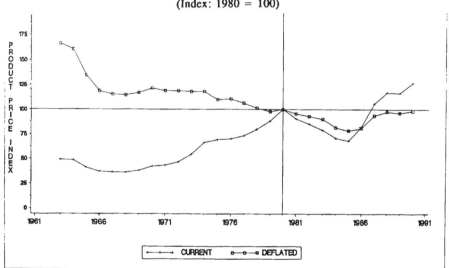

Source: As for Chart 3.

133

FARM FORESTRY AND THE ENVIRONMENT

Alberto Madrigal Collazo, Professor
Technical High School of Forest Engineers, Madrid

INTRODUCTION

Before analysing the consequences liable to result from expanding forested areas at the expense of agricultural and grazing land, consideration should be given to the following:

-- the links between agriculture and the environment;

-- possible uses of forests and the forestry techniques involved;

-- the layout of the area involved and the planned time-scale of the land-use change.

Subsequently, a clearer view should be obtained of the following questions:

-- the effects of land-use changes on the natural and socio-cultural environment;

-- possible alternative uses;

-- the implications for environment policy.

The links between agriculture and the environment are complex. Both in OECD countries and others, a wide range of types of agriculture - or, to be more precise, agricultural technologies - may be seen, with differing levels of environmental effects. At one extreme we find traditional types of agriculture, mainly in those countries with a long agricultural history; in such cases, extensive agricultural and stock-farming techniques are generally employed, implying a degree of equilibrium with the natural and social environment which tends to restore the fertility of the lands, favours the conservation of valuable forest wildlife and vegetation, promotes variety in the landscape and is a factor conducive to the stability of the rural population.

The other extreme is represented by intensive and specialized types of agriculture. These techniques, designed to increase productivity, involve high energy consumption due to the use of fertilizers and heavy machinery, and entail a series of negative environmental effects (OECD, 1991: Sumpsi, 1993) as follows:

-- as regards the soil: chemical contamination caused by fertilisers and pesticides; compaction caused by the use of heavy machinery; erosion and transport of solid materials resulting in the silting-up of reservoirs; salination caused by excessive irrigation;

-- with regard to water; contamination by fertilizers and pesticides of surface and underground waters; eutrophication of rivers and reservoirs;

-- with regard to the air, pollution is caused in particular by the difficulty in disposing of manure produced by intensive livestock farming (without grazing land);

-- with regard to animal/plant-life biocoenoses, since the use of fertilizers, pesticides and machinery calls for large areas to be brought under cultivation. This leads to a concentration of holdings and entails the ploughing-up of grazing lands and wooded areas, with the consequent destruction of hedges, borders and areas under natural vegetation, which provide the habitat for many animal species;

-- with regard to the landscape, which generally becomes monotonous as a result of specialization and single-crop farming, with its strong visual impact (e.g. huge silos and buildings to house livestock);

-- as regards the foodstuffs produced, which may contain undesirable quantities of the chemical by-products of fertilizers and pesticides, with negative effects for consumers;

-- as regards the stability of the rural population, since the use of machinery results in unemployment and can lead to emigration.

The forms of contamination referred to above have for some time been a constant concern of both public opinion and political leaders, and since the 1980s environmental policy measures have been developed in many countries, integrated in many cases with agricultural policy (Sumpsi, 1993). Nevertheless, it should be noted that in general agricultural pollution is widespread and the extent of the flows of contaminants is difficult and costly to evaluate. As a result, difficulties also arise in the adoption of corrective measures and the identification of those farmers responsible for pollution (Sumpsi, 1993).

In between the two extreme cases referred to, an entire range of intermediate types exists with the result that diagnosing their effects and formulating corrective measures becomes extraordinarily complicated.

It is important, however, not to lose sight of the essential point: intensive farming techniques are inherently damaging to the environment, while extensive techniques are always favourable. By regulating the use of fertilizers, pesticides, machinery and irrigation systems, the harmful effects of intensive agriculture can be held to within acceptable levels; in many cases, extensive agriculture and livestock farming can, on the other hand, lead to situations which have serious effects: e.g. cultivation of marginal lands on steep slopes or overgrazing (danger of erosion).

In principle, it might be thought that the cessation or reduction of intensive farming would eliminate the negative effects referred to; however, some of those effects are liable to be more or less durable (e.g. the silting-up or eutrophication of reservoirs, the presence of heavy metals in the soil).

For many centuries forests have been used almost exclusively for the production of firewood and timber and to a lesser extent for pasturing, game and other products such as cork, resins and fruits, which were of lesser economic importance.

The perception that forest resources - principally timber and firewood - were scarce and could be renewed led to the development of silviculture techniques. Through a felling and replanting sequence, these techniques ensured that forest resources were renewed; their incorporation into forest management has made it possible to bring tree populations close to equilibrium as regards distribution by age or size, guaranteeing a sustainable yield. These provisions are aimed at optimizing silviculture systems and felling ages to achieve the maximum product yields.

This rational and balanced system, aimed at renewability and the maximum sustainable yield of forest products, is embodied in the technical rules contained in the forest management provisions in effect for a century or more in Central Europe and the Mediterranean countries (Portugal, Spain, Italy and Turkey). These provisions are applied by the forestry authorities in public and State-owned woods, and to varying degrees in private woods. The management of privately-owned woods varies greatly from one country to another, with some countries applying the same rules as to public woods, while others merely exercise some degree of control; in yet others no restrictions are imposed on tree-felling apart from the imposition of taxes on the products obtained.

In many private woods in OECD countries and others selective felling is practised, which involves harvesting the best individual trees and does not conduce either to renewability or a sustainable yield. This practice was - and still is - widespread in low productivity forests in Mediterranean areas; combined with frequent periods of overgrazing, it results in the serious degradation of the lands concerned.

The following presents a broad picture of the forestry situation in the OECD countries:

--	In northern countries and regions (Finland, Sweden, Canada, United States) with relatively smooth topography and vast natural conifer forests of the taiga type (few varieties, reliant on sunlight, suitable for cool climates), a simplified form of forestry is practised, based on clear felling over large areas followed by artificial regeneration. In general, these forests give a high timber yield and their management is aimed at optimizing production and minimizing operating costs. In many of these countries genetic improvement programmes, designed to obtain better yields in both quantitative and qualitative terms, have been developed over a number of years. This type of forestry, which is concerned almost exclusively with timber production, supports some large-scale industries which in many cases export their production (sawmills, wood processing, cellulose, paper); the expression "single-purpose forestry" was proposed by Schütz (1990) to describe it. According to the same author, the past few years have seen an attempt to ease the rigid forestry techniques employed by reducing the areas felled, introducing a mixture of varieties when replanting, and changing over to shelterwood management (phased felling).

--	In certain OECD countries and regions intensive afforestation programmes have been implemented, aimed exclusively at timber production and using fast-growing varieties, both native and imported. Man-made forests have been planted, consisting of *Spruce* and *Pseudotsuga menziessi* in Northern and Central Europe, *Pinus radiata* in New Zealand, Australia and Spain, and *Pinus pinaster (maritime varieties)* and *Eucalyptus* in Portugal and Spain. The forestry techniques employed are also of the "single-purpose" type, since clear felling is practised followed by artificial or natural regeneration (Eucalyptus). In certain cases the small scale of the holdings makes them unsuitable for any other forestry system, while in others this system is used to reduce operating costs. These artificial forests also form the basis for large-scale industries.

-- Substantial areas of forest - mainly public - in Central and Western Europe, including Mediterranean areas, have been subject to regulation for some decades. The forestry techniques used are as described above and are aimed at ensuring renewability; natural regeneration is augmented by artificial regeneration to achieve a sustainable yield. The tools available to the administrators have either been perfected (e.g. inventorying techniques, production charts), or are under development - which is proceeding satisfactorily (genetic improvement programmes). In addition, it is generally believed that the balanced structure of age and size categories in these forests which guarantees a sustainable yield, also gives rise to a fairly steady yield of non-timber products (e.g. game) and enables the lands to be used for social and recreational purposes (Schütz, 1990). As the latter author remarks, attempts are being made to formulate a multi-purpose forestry system, designed to embrace all possible uses of forest lands without prioritizing timber production. Some steps towards this end taken in recent years include the planting of vegetation to supplement natural regeneration (to obtain diversity of flora); the dispersal and reduction in size of areas to be felled and replanted, to improve soil protection and mitigate the effects on the landscape; and experiments with multi-criteria optimization models.

-- In the Mediterranean areas (large parts of Portugal, Spain and Italy; the South of France; Greece and Turkey), while continuing the production methods described above, an attempt is being made to formulate a Mediterranean forestry system, which will perhaps be little more than a difficult compromise between Central European forestry techniques and the characteristics of Mediterranean forest ecosystems: heterogeneity and instability (Quezel, 1977), low profitability of products and significant positive externalities (protection) (Madrigal, 1992). Although moves to arrive at a synthesis between these two elements are being made, they are dispersed and isolated. Perhaps all that can be done is the development of forestry techniques designed to prevent forest fires; in many countries preventive measures are not strictly enforced in view of their high cost and the scanty returns (in terms of products).

The outline given above, although incomplete, may enable an indication to be given of the alternative uses made possible by expanding in the areas under forest, and the resulting environmental effects.

The geographical structure of areas involved in land-use changes largely determines their effects on the environment and also on policies. Large contiguous areas present a very different problem to small, discontinuous parcels, although the same total area may be involved. Similar considerations arise as regards the time-frame for the project: if the change is to be effected over a short term, the effects on the natural and social environment may be drastic; by contrast, if the project takes place over a longer term the negative effects will be mitigated.

Both the time-frame and the spatial structure of the project will be conditioned by a number of factors. For example, where publicly-owned lands or large private estates are involved, preference may be given to short-term programmes relating to extensive contiguous areas. By contrast, where smallholdings are involved, land-use changes to small, discontinuous areas may be phased-in over time - except in cases where the lands have been abandoned due to emigration. In addition, land-use changes will be affected by the density of the rural population: where the density is high, programmes must be carried out prudently and phased in to ensure that displaced populations are rehoused. These restrictions will favour medium-term programmes carried out over small areas, if the intention is to subsidize - or provide incentives for - the change of use.

Finally, although this may be to anticipate one of the conclusions to this document, it should be stressed that land-use changes relating to small, discontinuous areas, carried out over the medium term, appear to be the most rational solution if the ecological and socio-economic effects are to be minimized.

ECOLOGICAL IMPACT OF CHANGES IN FORESTRY USE

The effects of changes in forestry use need to be considered in terms of the system or ecosystem overall, as its component elements are interdependent (biotope/biocoenosis). However, this paper examines the impact on each of these elements - climate, land, hydrological cycle and biocoenosis - in turn, as each will respond differently to the change in use and will be affected in different degrees.

Climate

Enlarging the surface area forested does not cause any major change in rainfall patterns or temperatures, except where such areas are really vast (thousands of kilometres). The existence of new forests will not increase rainfall but will increase condensation (dew, frost, etc.), which in turn will increase the input of water to the system.

Like deforestation, however, afforestation will cause major changes in the microclimate: it interferes with solar radiation and light; moderates sudden changes in temperature; maintains atmospheric and soil humidity levels almost constant; reduces wind speed and noise. This latter characteristic has been put to good use down the ages in planting trees as windbreaks to shelter crops, or more recently, as noise barriers along motorways.

The impact of forests with respect to climate change is widely acknowledged to be beneficial as the forest canopy traps CO_2. The larger the forest and the greater the number of trunks and branches and the speed of lignification (tree and cover growth), the more pronounced is this effect.

Nevertheless, these positive effects should be viewed with some caution. In the first place a forest microclimate will develop only once the cover has matured sufficiently, measurable by the crowns cover, and by the height of the trees. Secondly, forest use could also have a negative impact on the microclimatic conditions created. The felling of perimeter trees and sudden clearings in the canopy as a result of regeneration felling change the conditions prevailing in the forest interior. Shelterwood felling could minimise this effect. In -situ burning of felling and pruning debris would result in free migration of the stored CO_2. Without such maintenance (no intervention) the slow rotting of trunks and branches tends to keep CO_2 in balance. Burning timber rapidly releases the CO_2 that has been trapped for decades to the atmosphere.

Soil

Forest soil owes its fertility levels to the maintenance of a balance between it and the trees and other vegetation: continuous source of nutrients from leaves, fruits, branches, etc.; effect of root systems at deeper levels; entrapment of atmospheric nitrogen. The vegetation cover itself protects this equilibrium against the depredations of the climate: topsoil or subsoil erosion by rain; denudation and transport by wind; deterioration of organic matter by radiation and heat.

Agricultural soils on former forests or natural pastureland will of course have very different characteristics to the original soil. The biogeochemical cycles will have altered radically, due to the disappearance of ground cover and the absence of tree root systems at deeper levels. Agricultural techniques attempt to maintain fertility by tilling (to bring nutrients at the deeper levels, which crops cannot reach, up to the surface), by adding organic or chemical fertilisers (intensive farming) and by letting land lie fallow (extensive farming). Much of the benefits of the latter are reduced or lost by stubble burning, a very frequent practice in Mediterranean areas.

Agricultural soils generally have little organic matter in the upper horizons, due to the removal of vast quantities of biomass (annual harvest) with virtually no replacement, to the rapid mineralisation of the few organic residues which do not have the benefit of a forest micro-climate and, often, to stubble burning (Gandullo and Serrada, 1993).

While not wishing to present an exhaustive typology, the characteristics of cultivated soils in the OECD countries can be categorized broadly as follows:

-- Siliceous or limestone soils which do not have impermeable lower horizons of accumulated clay or calcareous rock.

-- The absence of such horizons is a characteristic of dry farming soils in areas where the parent material has few alterable minerals or in high relief terrain as is invariably the case in Mediterranean climates (Gandullo and Serrada, 1993) or in cereal farming soils where the characteristics of the parent material are the same, but the climate is cold and wet (no summer dry season).

-- Siliceous soils with a compacted horizon of accumulated clays characteristic of Mediterranean soils on alterable parent material: shale, modern sedimentary rock, etc. (Gandullo and Serrada, 1993). Irrigation of these soils aggravates the problem of the impermeable horizons.

-- Limestone soils which have calcareous rock horizons are also typical of the Mediterranean regions.

-- Calco-siliceous soils with calcareous rock horizons overlying a horizon of accumulated clays. Characteristic of Mediterranean areas, irrigation accentuates the problem of the impermeable layers (Gandullo and Serrada, 1993).

-- Soils which exhibit hydromorphic phenomena in the lower horizons, characteristic of wet climates and parent material with a clayey or silty texture or sediment with a high gravel content (Gandullo and Serrada, 1993).

-- Prairie or steppe soil, in humid climates, without a prominent horizon of accumulated clay or calcareous rock, due to the climate and the continuous presence of the grass cover.

-- Laterized soils characteristic of tropical climates, not frequently occurring in OECD countries.

-- Soils that have been salinized by irrigation.

How would afforestation affect these soils? Once trees are planted and have become established we might expect to see the following:

-- Topsoils enrichment by organic matter;

-- Appreciably reduced clays and calcic salts migration at deeper levels;

-- Root activity in the subsoil.

Finally, we would expect to see the beginnings of a slow but positive soil formation.

Actually, the real problem is not the effect of the new forest population on the soil, but the difficulties in planting a new tree population due to the soil structure. In some cases is may be possible to overcome such difficulties, partly by selecting suitable species (whose root systems will be capable of withstanding clay horizons or calcic salts) and partly by employing soil preparation techniques (subsoiling techniques in the case of calcareous horizons: horizon inversion if there are accumulated clay horizons; ridging, etc.) If the impermeable layers are too deep for farm machinery (ploughs, rippers, etc.) the problem will not be easily resolved and there is little chance that the forest will be viable.

There will still be some risk of soil impoverishment, for example, acidification of the topsoil by certain coniferous species in soils with a low pH in cold climates, which could trigger podsolization.
Changing the species and/or employing a mixture of broadleaf trees and conifers could be a successful solution to problems of this kind.

The reduction of fertility or soil exhaustion which results from the use of fast growing-species (specifically species of the Eucalyptus family) should be carefully studied. In a study of Eucalyptus groves in the North of Spain, Bara and al. (1985) demonstrated that there was no significant difference between soils under oak groves, pine groves and mature eucalyptus groves of over 30 years old. Soil exhaustion in Eucalyptus forests may be attributable to the short rotation (9, 10, 12 years) commonly practised in such forests. In such a short period the forest population cannot replace all the nutrients that it has taken from the soil.

This phenomenon, which is also observable in other plantations (poplar) and in coppice broadleaf forests subject to short felling cycles as well as Eucalyptus forests, could be remedied by extending the rotation or felling age, in an attempt to reconcile or strike a balance between ecological and economic interests. Intensive forestry, which uses fertilisers and genetic improvements, may offer another solution. This type of forestry, for timber production, is analogous to intensive farming, although it uses less fertiliser as a rule and the harvest is not annual.

Generally, efficient forestry maintains the soil/vegetation balance, and thence fertility, throughout the life-cycle of the forest population. Only regeneration felling can change this situation, the more it is done, the larger the breaks in the canopy produced by felling. Clearly, this type of change triggers the establishment of a new balance as natural or artificial regeneration begins to occur.

Soil preparation prior to planting can also introduce other potentially negative factors, especially in mature soils. This is not the case with afforestation of agricultural land where, as we have said, the soil has been impoverished. In this case, the negative effects could result from the application of techniques which bring materials containing lime and chalk to the surface.

Erosion and the hydrological cycle

Erosion, the process by which surface horizons are partially or completely worn away, affects all soils to a greater or lesser extent, except where there is 90% vegetation cover (Ruiz de la Torre, 1991). In cultivated soils, the absence of permanent vegetation cover and its annual removal at harvest time inevitably results in erosion. On plains or gentle gradients, erosion takes the form of vertical migration of the upper horizons and may even expose horizons of accumulated clay initially formed in lower layers, causing total denudation of the surface. In mountainous terrain transportation can occur, leading in extreme cases to the total removal of the topsoil and the formation of rills and gullies.

Water erosion is also influenced by other factors, such as rainfall pattern, terrain relief and type of parent material. Deposition and sedimentation can accompany erosion and transportation and, where they are intense over a short period, can have a very negative impact on other crops and reservoirs (silting up, etc.)

Wind erosion causes soil loss through transportation and may occur either on its own or in conjunction with water erosion.

Water and wind erosion are both factors for all cultivated soil, but not for grassland or pastureland which is protected by dense grass cover.

However, on steeper gradients on both land under crops and pasture there is an imbalance in the hydrological cycle. Rain-water runs off the slopes with only a little infiltrating into the soil, and evaporates to a greater or lesser extent, because there is no, or only low, ground cover. In contrast, dense tree cover (including shrubs and tall brushwood) regulates the hydrological cycle. It is well established that the complex structure of crowns and branches of trees and shrubs breaks the fall of raindrops, allowing the rain to fall gently to the soil (thus preventing denudation, which initiates erosion). This, together with decaying waste (fallen leaves) aids infiltration. The canopy prevents or reduces evaporation.

To sum up, any dense vegetation cover (grass, scrub, shrubs, trees) reduces or inhibits erosion; but tall cover (trees, shrubs and scrub) is more effective at regulating the hydrological cycle. The protection afforded by forest and shrub cover is extremely important in mountainous and Mediterranean areas, where they also reduce the flood damage caused by torrential rain. Rain-water runs off down the slopes, without transporting any material that would exacerbate damage to crops, reservoirs, transport infrastructure, homes and even human life.

Gallery and riparian forests also protect riverbanks from erosion by rivers and, where they replace intensive crop farming, avoid contamination of river water by fertilisers and pesticides.

It is clear from the foregoing that, in general, converting to forestry use prevents erosion and improves water quantity and quality.

The silvicultural practice must take due account of the risks of erosion and changes to the hydrological cycle when selecting methods of regeneration felling.

Where there is a risk of erosion it should prohibit clear cutting and use only regeneration felling (shelterwood felling) or patch logging. It must adhere to the felling strips and avoid felling along watersheds or the banks of rivers, flood channels and streams.

In any case, the critical factors as far as erosion is concerned are the use of heavy machinery in logging operations and the construction of skidding roads in mountainous areas. There is a growing awareness of such problems and in many countries cables or small tractors are used for logging operations as they have less impact on the soil.

Biocoenosis

The plant life biocoenoses in agricultural ecosystems are poor, infrequent, change radically with every harvest and only achieve some degree of stability on the boundaries or periphery of crops. Conversely, on grassland and pastureland there is a high degree of biodiversity, although its stability is dependent on grazing by domestic or forest animals) without which these grassland systems would become overgrown and regress.

Just as the abandonment of grazing would cause regression, the abandonment of crop farming would make room for potentially rapid natural colonisation, initially for smaller plants of no great ecological importance (annuals). This phase would gradually be succeeded by others although development would be extremely slow. This slow progression might be halted by any major disruption (fires in dry regions of the Mediterranean. The planting of a tree population on some land under crops would speed up certain phases.

The characteristics of agricultural land will dictate the choice of species to be planted. They will generally be hardy (capable of surviving in poor soils) and light-tolerant species (capable of establishing themselves without protective cover). This would mean starting with monospecies populations of the same age: factors which do not help stability. But the experience gained with past reafforestation schemes suggests that other, shade-loving, species (which need the shade to grow) will slowly become established under the cover of the monospecies population. This will create the conditions for a diversification phase that will allow the forest population planted to become stable.

Before converting grassland and pastureland to forest, we must consider whether:

-- it is land that has permanent plant communities with climatic and soil conditions that are unsuitable for forests (mountain pastures, grass steppes, etc.)

-- it is land that been cleared by ploughing up forests for use, more often than not, for extensive livestock farming.

In the first case, the advisability of converting to forest use is highly questionable and the ecological impact will invariably be negative. A diversified herbaceous community would be replaced by a uniform arboreal community.

In the second case, conversion may bring positive results, although the first phases of afforestation will reduce the initial diversity. Where conditions are favourable (Mediterranean climate, relatively flat relief) the most effective solution may be to opt for the type of wooded pastureland which is typical of the Mediterranean forestscape, and a genuine silvopastoral system or ecosystem. (The last example assumes that the maintenance of extensive livestock farming or hunting use is appropriate).

Animal biocoenosis is directly dependent upon plant life biocoenosis, since any change in the latter will have a major impact on the former. The microfauna can change drastically if the vegetation detritus which is its food changes but the organic horizons which are its habitat are unchanged. A perfect example is what happens when pastureland is forested: the earthworm population, which actively transforms the soil of the pastureland will be replaced by microarthropods which will begin colonising the changed ecological niches.

The macrofauna is more adaptable to change. Some species can even change their feeding habits. There is less species diversity than among the microfauna and some species may disappear from an area altogether or their numbers may decrease as a result of the change to forest use. This may happen with grain and fruit-eating steppe birds which have adapted to fallow land and land under crops. Other species, on the other hand, may expand their territory to the shelter of the new forest areas.

If the conversion to forest use operates over areas that are not contiguous, it will form a mosaic of wooded areas, crops and pastureland which would be very beneficial for fauna, allowing them to shelter and breed in woodland and feed of crops and pastureland.

Moreover, the age balance that cultivation techniques seek to achieve as an integral part of Forestry Management, in order to ensure a sustainable product yield, prefers the same scenarios for both protected wildlife and game.

As regards biocoenosis, two aspects of forestry management are critical. First, the need to preserve the habitat of protected fauna; this can be achieved through a series of conservation measures which go from timing logging operations so that they will not coincide with the breeding season, to a ban on such operations (establishing reserves). The second, relates to the fact that many regeneration felling methods lead to the establishment of monospecies forests. The species enrichment plantations previously mentioned can do much to correct this.

EXTENSION OF THE FOREST AREA: ALTERNATIVE USES

Suitability of forest land and demand for various uses

The first point we have to address in outlining a survey on the future use of forested farmland, would be to examine the extent to which new forests would be suitable for generating, conserving and catering for social uses, and in what combinations. In short, to determine how new forests could fulfil a multi-purpose role. This implies that the species used in afforestation will be those most suited to the biotopic conditions and, consequently, that the forests will be viable. However there is still one question mark: can these artificial forests be considered forest ecosystems and therefore capable of fulfilling this multi-purpose role? Or are they simply plantations, and thence inherently monocrop systems?

The second point is to determine the extent of demand for these new resources. This, in turn, raises several important issues as to the origin of demand: what will be the level of demand from the rural population which is living in the immediate area and from the urban population? The urban population, which will always be a factor (greater awareness with regard to ecology and nature conservation issues), could influence the new forests to a greater or lesser extent (strong demand for green belts for recreational pursuits in areas close to the large urban centres, decreasing as the distance from the city increases).

Lastly, we have to review the effects of such change on land use. If it has been implemented over large contiguous areas, the afforested area will be managed in line with forestry policy principles. It will be up to Forestry Management to decide on the permitted uses, restrictions and cultivation methods.

If, on the other hand, the change has taken place over somewhat smaller, more fragmented areas, use becomes an entirely different problem. Agricultural and livestock uses, though theoretically reduced, will persist and will have to be integrated with other uses of the forest.

These two potential scenarios of -- large-scale or small-scale afforestation -- may well have important consequences for the stability of the rural population. Large-scale afforestation will automatically mean the end of crop and livestock farming, and although the surplus labour could be employed in afforestation operations, once these were complete, the rural population or a sizeable part thereof would have to migrate. Phasing in large-scale afforestation operations could lessen the impact of these negative trends as, with a more gradual change of use, the labour force would switch slowly from agriculture to forestry and its stability, as such, would be ensured since, once planting was complete it would be time to begin cultivation operations in the areas first planted. In this event, there would be no total cessation of farming and livestock rearing activities; afforestation operations could take place at times when there is not much agricultural activity and, in the future, both the availability of forestry work and income diversity would be stabilising factors for the rural population.

Problems relating to man-made forests and their potential uses

Above, we raised the question of whether or not man-made forests could be considered as forest ecosystems. Certainly, they behave in the first few years as crops or monocrops, but as time goes on their constituent elements (trees) are gradually welded into a single integrated system. We see the beginnings of competition in natural pruning, crowns differentiation in the overstorey and understorey and natural mortality (as trees are submerged). Interaction with the biotope begins: surface run-off ceases or is reduced, pH values and carbon/nitrogen ratios change; in short, a slow soil regeneration begins. As the canopy closes a forest microclimate is created, aided by the appearance of fungi, the establishment of shade-loving plant species and the disappearance of species that need more light (Madrigal, 1994). All of these natural processes can and must be controlled by forestry techniques. As one of the recommendations in the World Forestry Congress (Paris, 1991) states "Management of plantations should be so planned that silvicultural operations, particularly tending will ensure the transformation of a plantation into forests." In other words, the long process of systemic integration can be speeded up by proper silvicultural management, which seeks to achieve stability of the forest population. In this way man-made forests will be in a position to fulfil the multi-purpose role of natural forests.

But, not all man-made forests can follow this path to systemic integration and transform themselves into real forest ecosystems. To achieve this, competition between trees has to be the predominant relationship, at least when the forests are young or at the intermediate stage. Crown density, (and hence tree density) also has to be sufficient for the interactions with the soil to take place, and to establish a microclimate. The shape and size of man-made forests and the planting density are the main factors which determine whether interaction will take place or not. So, in linear plantations and narrow windbreaks, the main competition takes place around the long edges, with vegetation that is not part of the forest. At the Symposium on Man-Made Forests in Canberra (FAO, 1967), it was proposed that linear and windbreak plantings should not be considered as man-made forests. The proposed minimum size for man-made forests was at least 100 metres wide. In the past, the European Forestry Commission defined these small areas as plantations outside forests, when interspersed with crops and consequently associated more with agricultural income at the site.

When planting density is low, the trees have much more space to develop their root systems and crowns, which scarcely touch. The trees also have sufficient nutrients, water and light and there is therefore little or no competition between them. The free spaces are taken up by other plant species which take advantage of the nutrients supplied by the trees, and either take advantage of, or flee, the shadow they offer. The microclimatic and soil formation conditions are very different from those in a closed wood. In the way it functions, this type of low-density plantation is similar to the "dehesas" (wooded pasturelands) which are typical of much of the forestscape of Portugal and Spain. These can be regarded more as silvopastoral systems or ecosystems or even agro-silvopastoral systems, which cater for a combination of grazing, hunting and forest uses (fruit, cork and firewood) including agricultural purposes (tilled wooded pastureland). They also provide a suitable habitat for protected wild-life.

The problem in defining the density limits, is that they vary greatly with different species or groups of species. On Spanish wooded pasturelands, the density is expressed as a fraction of the cover capacity or crown spread and ranges from 5 to 20 per cent.

Given the foregoing, a classification of plantations is proposed, which enables a preliminary definition of their uses. The proposed types are:

-- *man-made forests,* whose size, shape and density will facilitate their transformation into real forests.

-- *off-forest plantations,* which from a usage point of view must be regarded as part of whole or a mosaic of crops and wooded areas.

-- *low-density plantations*, whatever the surface area that, with certain reservations, can be used simultaneously for agriculture, livestock farming and forest uses.
-- *windbreaks,* planted essentially to provide shelter.

-- *linear plantations,* generally ornamental.

148

In *man-made forests,* the planting and establishment phase rules out any use in the early years. This said, the forest population will already be protecting the soil and regulating the hydrological cycle at this stage, and will be able to do so fully once the canopy has closed over. From the shelter of its microclimate, wildlife will be breeding and slowly colonising the forest, especially around its fringes. In the medium term, certain uses will be compatible, if properly regulated, with this protective role. Light grazing by suitable livestock could be the first such use. In Mediterranean areas light grazing helps prevent forest fires, by keeping pasture short. There may be a lot of edible fungi within just a few years, which could provide a new economic activity. If the quantities are surplus to the requirements of the rural population they may be sold outside the immediate region or grown on an industrial scale. Game will be able to thrive, and may need to be controlled to avoid damaging young stands (rabbit, deer, boar).

It will take longer to introduce recreational uses, depending both on the facilities the region has to offer and on the growth patterns of the species planted. The landscape may be monotonous at first (use of one or only a few species, of small size) and will remain this way in the case of large-scale afforestation. Forestry techniques could plan for a mix of species from the start, provided that the conditions are right, or could combine the first fellings with improvement plantings.

Forest production, particularly timber production, would require a longer period still. Timber could only be taken after the first thinnings and then not in any great quantity until the 20 to 25 year mark, as a rule.

The timescale may be shorter for fast-growing species (Eucalyptus, poplars, some pines). For artificial forests the problem with productive use is in achieving sustainable yield. The much cited balanced age distribution, on which sustainable yield depends, is very difficult to achieve with man-made forests, since it has to start off with a population of the same age or within a very narrow age range. If afforestation has been phased in this will not be major problem but, in any event, fellings will either have to be brought forward or postponed, with respect to optimum age, in order to achieve a better age distribution in the forest. This, in conjunction with species diversification (for aesthetic purposes) could create the right conditions for multiple use and sustainable timber yield, although it would involve sacrificing some, possibly substantial, returns.

In *off-forest plantations,* for each and every small-scale plantation, a series of potential uses similar to those for man-made forests could be developed with the following provisos:

-- The edge effects, being more pronounced, could lead to an increase in wildlife which, while it could have a positive impact (wildlife protection or hunting) might also have a negative impact (damage) to neighbouring crops and the forest population itself.

-- The smaller size would prohibit the introduction of recreational facilities (picnic and camping areas) although in the area as a whole, once sufficient time has elapsed for the trees to reach an adequate age and size, other recreational activities such as hiking and horse-riding could be developed.

-- Forest production would be severely restricted, especially in terms of the sustainable yield objective, which would be impossible for individual small plantations, difficult for all plantations on a single property, but perhaps possible on a regional scale.

149

The mosaic of crops and wooded areas offers a rich and varied range of forest, agricultural and livestock uses, but these require planning and organisation, given the different interactions which develop between crops and tree populations. The restrictions which should be imposed on use will depend on whether priority is to be given to agricultural or to forest use. This, in turn, depends on the productive potential of the various uses, the interests of owners and policy-makers who decide whether to encourage or restrict use. As can be seen, there can be many variations on the theme. Whatever the case, the following precautions would be advisable:

-- Prohibition of, or tight controls on, the use of forest species which take a lot of water (Eucalyptus), because of the detrimental effect on groundwater used for irrigation and for human consumption.

-- Prohibition of, or tight controls on, the use of invasive forest species (Ailanthus, certain Acacias) which can begin to compete fiercely with crops as they attempt to colonise them.

-- Prohibition of, or tight controls on, stubble burning, the burning of harvest debris and forest vine and olive prunings in the vicinity of forests.

-- Banning livestock from forested areas until the trees are tall enough and old enough not to be damaged by the animals.

As we have seen, a mosaic of crops and wooded areas can be put to multiple (agro-silvopastoral) use, given certain restrictions and precautions, once sufficient time has elapsed. In the long term, forest use may give rise to the problem of renewability. The small surface area of each forest plantation, taken individually, would either rule out management by natural regeneration, or would require supplementing by artificial regeneration. Mature trees would have to be felled at the same time (clear felling). In practice, the process would be somewhat similar to the single-purpose forestry previously mentioned. For all the plantations on the same property, indeed, in an entire region, it would be possible to introduce a phased regeneration felling sequence, similar to a block felling sequence, which would be closer to multipurpose forestry and to the sustained yield objective referred to above. Clearly, considering off-forest plantations as a whole calls for forestry management plans and, consequently, for regulation of uses and products. If no such plans are made, owners may find themselves at a disadvantage when it comes to selling their wood products in the future (dispersed supply, depressed prices).

In addition, alternating plots under crops, pasture and off-forest plantations will, in time, make for a varied rural landscape from both a visual and an environmental standpoint (Gomez Mendoza, 1993). This could provide a framework for rural tourism in the sense of *farm tourism,* as a rural activity -- offering accommodation in *working farms, holiday cottages, guesthouses, farm camping, bed and breakfast, etc.* providing closer contact with country life (Cals and al., 1993) -- or for *rural tourism in the strict sense of the term,* which with the same end in view, uses accommodation which would have a limited visual impact on the countryside (country inns, hotels, camp sites). These uses, which may be described variously as *green tourism,* environmentally-friendly tourism or alternative tourism (Cals and al., 1993) promote stability and development of the rural population and its development through the jobs it creates and the income it brings in. It would also entail few negative externalities.

Low-density plantations would also need some time before they could be put to silvopastoral uses, as it would be necessary to protect young trees from grazing livestock. Moreover, if trees are planted on land that was formerly cultivated, it would be necessary to establish artificial pasture and wait the requisite time before it could be used, after which -- the length of time would depend on the type of pasture and type of tree, on whether or not tree-guards are used, and on the type of livestock -- livestock could be allowed to graze; or hunting and intercropping, if the aim is agroforestry use.

In this case, a special kind of forestry would be required. The work necessary to strengthen the plantations and eventually obtain products from them, would be on a tree by tree basis. Rather than forestry techniques, which are applied forest-wide, the need would be for arboricultural techniques. Products depend on the tree type: timber, from the plantations of *Pinus radiata* which have been planted for some time now in New Zealand and other countries; top quality timber, from *Juglans* ; cork, fruit and firewood, from *Quercus suber* in plantations in Portugal and Spain; fruit and timber, from *Castanea* and *Pinus pinea*. Only artificial regeneration could be used in this type of plantation, since natural regeneration would not succeed, given its use as grazing land, or for hunting or crop farming.

Once the trees have matured, the quality of the landscape will differ with the species planted, but will inevitably be monotonous in monospecies plantations.

Low-density plantations can be incorporated into mosaics of agricultural land and off-forest plantations thus offering a wider range of uses and a more varied landscape.

These mosaics of crops, off-forest and low-density plantations are examples of the multiple use of rural land. In English-speaking countries such use is termed "Agroforestry" or "Farm forestry", and "Foresterie paysanne" in French-speaking countries. Spanish as yet has no preferred term.

The last types of man-made plantation proposed, *windbreaks* and *linear plantations,* are established with a very specific purpose in mind which drastically limits potential use. Windbreaks are used to shelter crops from damage by prevailing winds, while linear plantations are used to provide shade and for aesthetic reasons. Both types can provide habitats for animal species, especially birds.

In conclusion, the range of uses of the various types of artificial plantations can be summed up as follows:

-- Man-made forests: multi-purpose forest use.

-- Mosaics of off-forest plantations and agricultural land (Agroforestry): agricultural and livestock uses and restricted forest use.

-- Low-density plantations: in isolation, silvopastoral, game and even silvo-agricultural use; or together with land under crops for agroforestry type uses.

-- Windbreaks and linear plantations: shelter and landscape feature, respectively.

In all cases, there will be some delay before the plantations can be fully utilised.

In the case of crops and off-forest plantations, a major additional use which can contribute considerably to rural development is briefly analysed below.

Rural tourism

As previously mentioned, rural tourism in both its forms -- accommodation provided by the local population (farm tourism) and low-impact purpose-built tourist accommodation -- is a further use related to agroforestry systems and the country sides they support which, because of its seasonal nature, is of increasing interest.

The development of these desirable subsidiary activities may be impeded by a number of factors. Some of these relate to the characteristics of the rural population (depopulation, demographical ageing, lack of social life, routines or even defensive tactics by the local population); others relate to the lack of infrastructure (inaccessibility of a region, local population lacking sufficient resources to bear the costs of fitting out accommodation, lack of cultural, social and commercial amenities); and, lastly, other factors specific to the rural environment, its deterioration for example (Cals and al. 1993)

In some cases these factors can be dealt with relatively easily: the deterioration of the rural environment can be halted by launching an afforestation drive and abandoning or reducing intensive crop-farming and livestock-farming activities; the costs of fitting out accommodation and of creating an infrastructure of small hotels are not very high. The negative factors relating to the characteristics of the rural population are the most difficult to remedy.

The sale of handicrafts could boost rural tourism, making it more attractive for the user and providing a new source of income for local inhabitants.

This type of tourism is not mass tourism and would produce little or no negative externalities, while the economic benefits would stay within the region. The use of rural accommodation could facilitate other more conventional tourist activities (winter sports, hunting) or the development new activities for which demand is high (cycling holidays, adventure sports such as rafting, mountain-biking, hang-gliding, etc.). Such activities can be developed outside the immediate region; they tend towards mass tourism, can have a major negative impact and most of the profits go to promoters or proprietors rather than the rural population.

Another form of mass tourism, or on the way to becoming so, which generates negative externalities, is the promotion of trips to nature reserves (regional or national parks), taking advantage of the growing awareness of nature conservation issues. This activity could be part of the development of genuine rural tourism if it used its accommodation infrastructure. But, in the considered opinion of Cals and al. (1993) it is an activity which can be "at nature's expense - an intrusion - rather than an immersion in nature" compared to rural tourism which is an immersion in nature and treats it with respect.

Demand

When we speak of demand for products and use, in relation to man-made plantations, we are referring to future demand, given the long lead-times involved.

The demand for forest products, particularly timber, was thoroughly covered in panel two. Labour demand (and supply) was discussed in panel one.

It is assumed that the demand for agricultural and livestock products will have matched supply (it would be more correct to speak of supply adjusting to demand) once land has been taken out of agricultural use and surpluses have been eliminated.

However, we should still review briefly some aspects of demand for subsidiary uses generated by forested land. Such demand will principally be from the urban population, which is generally aware of issues such as contamination and nature conservation. Moreover, there is a marked and very high demand for green spaces for leisure activities which the recreational amenities the new plantations offer could satisfy. The demand for farm tourism and rural tourism is increasing or has reached acceptable levels in many countries of the OECD (Germany, France and the United Kingdom, among others). In other countries the demand is just beginning, perhaps largely due to the fact that there has only been an urban population for some twenty or thirty years and its ties with its rural roots are still strong.

Demand for open-air sports, hunting and trips to nature reserves is also increasing and could use the rural tourism accommodation services, as mentioned in the previous section.

The relationships between the supply of such services and demand for them call for policies which will promote rural tourism activities and regulate activities which can give rise to negative externalities (mass tourism).

Spatial aspects of the change in forestry use

Throughout this paper, we have repeatedly referred to two options: large-scale and small-scale afforestation. Opting for either one or the other might well lead to inefficient allocation of use, at either end of the scale. The abandonment of fertile agricultural land could lead to total afforestation. The existence of a rural agricultural population on land of mediocre quality, leaves small-scale afforestation as the only alternative (afforestation of land subject to erosion, or producing surpluses).

From the foregoing we can conclude that the presence of a rural agricultural population means that the least drastic solution would be a change to agroforestry or agro-silvopastoral systems. The advantages of these systems are that they promote a well-cared for natural environment, as well as offering multiple services, including subsidiary services, which could boost sustainable rural development.

The last point I would draw attention to is that in both scenarios, gradual afforestation, would seem the most suitable method all round (in terms of the environment, supply sustainability, and in social terms).

POLICIES AND CHANGES IN FORESTRY USE

Afforestation as an instrument of policy

The afforestation of lands denuded of trees is and has been used as a means of accomplishing a variety of objectives. For example, in policies designed to control water resources, the afforestation of catchment basins has been found to be an important instrument in protecting soils against erosion, regulating water runoffs, preventing the silting up of reservoirs, preventing floods, etc. Of course, afforestation alone cannot achieve these aims, and must be implemented in conjunction with other measures (regulation of crop cultivation, construction of dams, dikes, etc). Further, if the plantations in question are to fulfil their protective role suitable treatment must be applied on a continuing basis.

Comprehensive and well-designed forestry policies can also determine the increase in forested areas required to meet various aims: increase in timber production, regeneration of impaired ecosystems, social facilities. Follow-up treatment is also required vis-à-vis the new forested areas if they are to continue to achieve the aims desired. In these policies, however, afforestation is not the sole instrument but merely one aspect of forestry regulation, which includes protective programmes (against fire, acid rains, blights and diseases), programmes relating to product marketing, the preservation of habitats, recreational activities and hunting, the training of forestry workers, etc.

Specific forestry policies, such as those developed by the owners of timber processing plants, make use of the afforestation of lands denuded of trees to maintain continuous flows of raw materials. However, in this straightforward case in which afforestation is an essential element, management of the new tree population is also indispensable.

All the cases referred to have features in common, apart from afforestation: the long time-frame and the need for continued protection measures (hydrological policies); the wide variety of uses (global forestry policies); timber production (specific forestry policies). In addition, in the first two cases afforestation is not the only instrument available for achieving the objectives sought.

This preliminary discussion highlights the fact that the afforestation of agricultural lands, especially in the form of a mosaic of wooded and agricultural areas (agro-forestry), is not the only means of attaining the objectives of these policies, which should be viewed globally as agro-environmental policies. Above all it has drawn attention to the danger of over-simplified solutions arising from specific situations in which, once the immediate aim of the conversion to forest use has been established (elimination of agricultural surpluses, reduction of agricultural pollution) no consideration is given to other measures or the long time-frame involved.

The bases of agri-environmental policies

Agri-environmental policies should incorporate the objectives referred to above (elimination or reduction of agricultural surpluses and negative effects of farming) which are of varying importance in the OECD countries. However, these two objectives must be integrated with the more general aim of sustained rural development - which is obviously a long-term process. This entails the implementation of a series of programmes, some to be launched immediately - such as afforestation and perhaps rural tourism programmes - and others adapted to the pace at which new uses become possible and to the consequent demand.

The delineation of the geographical area concerned by the proposed rural development measures will influence the related policies and in turn be influenced by them. Some countries prefer to segregate land uses via rural development provisions: forested and agricultural lands are considered separately and the afforestation of agricultural lands would call for the reclassification of the latter as forested areas. In other countries, where marginal agricultural lands are abandoned, a similar solution may be adopted, as we have noted previously in this paper. In yet others, by contrast, land uses are not precisely defined so that the option exists of establishing a mosaic of integrated forest and agricultural areas. The separate classification system implies that the lands tend to become subject to forestry policies, whilst the integration of forested and agricultural lands implies the adoption of agro-environmental policies.

A key element in the success of these policies consists in establishing the closest possible link between the rural population in question and the programmes to be implemented. In many cases, the failure of forestry policies has been due to the non-existence - or breakdown - of links with the rural community (incompatibility of extensive livestock farming with forestry, for example). In framing environmental policies, integration should be regarded as indispensable; in general, it will be easier to achieve than under forestry policies since lands earmarked for continued agricultural and stock-farming activities - in addition to the newly forested lands and the accommodation facilities provided for rural tourism - are owned by local inhabitants. In any case, in the context of agro-environmental policies extreme care must be taken in adopting decisions regarding land-use changes to take proper account of the views of the landowners and also to avoid any possible tendency to see rural development as a closed circuit - leading in extreme cases to "Indian reservation" situations.

Another important aspect of these policies (and of all policies concerned with natural resources and the environment) relates to the externalities generated - in this case by agro-forestry systems. In formulating these policies, it would be helpful to stipulate that positive externalities (protection, conservation and - to some extent - provision for recreational uses) will be remunerated by the public authorities. In this way, public measures to protect non-marketable resources could be introduced, as recommended by Wibe (1992) in connection with forest management. On the other hand, positive private-sector externalities (recreation, which requires specific infrastructures, and other uses such as hunting and open-air sports) would be paid for, at least in part, by the users and promoters.

Agri-environmental policy measures

It might be thought presumptuous to propose a list of agri-environmental policy measures and instruments when various countries participating in this seminar have presented their agro-forestry programmes which are at various stages of implementation and development. Nevertheless, we thought it of value to propose for subsequent discussion a series of possible measures believed to be suitable for launching agri-environmental policies which include afforestation. We do not wish to assert that afforestation, whether total or partial, is the keystone of these policies; other measures exist (organic farming, for example) which could contribute to the solution of the problems under discussion.

The following are the measures we propose:

As regards afforestation and wooded areas

-- Measures designed to facilitate the implementation of specific afforestation projects, especially in those countries in which regulations are in effect governing the choice of varieties, the evaluation of the environmental impact of afforestation, etc. These measures could be used to establish the territorial limits - whether at regional level or throughout a homogeneous geographical area - in which the selection and evaluation criteria apply, avoiding the proliferation of micro-projects which could lead to disinterest and abstention on the part of owners of forest lands, or an impenetrable morass of administrative procedures.

-- Financial aid for afforestation projects. The majority of agricultural landowners are not in a position to finance the high costs of afforestation unaided, although they could, as their personal contribution, provide part of the necessary manpower. It would be appropriate for the public authorities to finance these costs in whole or in part. Such financing could be considered as remuneration for the positive externalities (protection) which will follow the establishment of the new forests. A variety of means could be used: tax concessions, contracts with the authorities, subsidies in the form of annuities or combinations of such measures. It would also be helpful if short-term loans on special terms could be granted to landowners intending to plant forests on part of their lands, to bridge the gap between the disbursements relating to forestry operations and the receipt of financial aid.

-- Financial aid for the cost of the initial treatment of the newly-established woods. To ensure the viability and stability of the plantations, treatment is required (treatment of competing vegetation, clearing, etc), calling for fairly substantial investments. These tasks, which may continue for 5-10 years, should be aided via the same machinery as for afforestation, since they are carried out in pursuit of the same ends.

-- Financial aid programmes to offset the loss of agricultural incomes until the newly-planted forests provide their first profitable yields (average time-scale: 20-25 years).

-- Technical support measures for afforestation and the initial operations, including tasks relating to extension work, vocational training, tree nurseries, treatment of diseases, etc.

-- Measures to regulate forestry production.

Aids similar to those referred to under the previous indents have been proposed in the context of the reform of the EU agricultural policy.

Relating to the harmonization of agricultural activities with the new forested areas

This series of measures could range from an ambitious plan for the regulation of crop cultivation to simple measures to regulate conflicts between different uses, such as the restrictions or prohibition applying to the burning-off of fallow lands, already referred to above, and harvest residues close to forested zones, or the temporary ban on grazing in the latter. Measures of this kind (agricultural codes of conduct) should be proposed by the public authorities and supported by information campaigns and extension work.

Relating to rural tourism and recreational activities

-- Aid measures for the fitting-out of accommodation facilities for agro-tourism. Aid should be given not only in the form of public subsidies and/or loans on special terms, but vocational training programmes (building trades, carpentry, catering trades) should be organized and measures taken in support of the production and promotion of crafts products. The EU LEADER programmes could provide a good example.

-- Measures to promote the creation or improvement of rural tourism infrastructures such as small hotels, and the infrastructures required for open-air leisure and sports activities (picnic areas, camping sites, hiking and cycling paths). The financing of these operations would be a matter for discussion, since the greater part should be provided from private sector sources (e.g. small firms, users).

-- Environmental effects. Many of these activities are - or should be - subject to specific regulations (winter sports, hunting); others, however, may require such regulation due to their growing popularity and the increasing numbers involved (cycling, adventure sports).

Many of these measures aimed at subsidiary activities call for coordination between different government agencies (agriculture, tourism, environment, vocational training) to ensure their efficient implementation.

Measures to promote the formation of associations

The formation of associations by owners of forest lands can protect the markets for a varied output of agricultural and forestry products from dispersed sources; especially in the case of forestry products, the marketing of fragmented amounts can depress prices. As regards forestry, other significant advantages can be obtained through the formation of associations: joint use of tree nurseries, preventive measures against disease and fire, timber production, joint use of access roads to plantations subject to regulation, to prevent uncontrolled proliferation of these infrastructures with their consequent negative effects.

In addition, associations formed by the owners of agro-tourism accommodation facilities can help to improve these services.

Finally, the formation of associations can promote and intensify extension, information and vocational training activities.

We believe that in combination, the measures referred to - though the list if by no means exhaustive - could contribute to balanced and sustainable rural development.

BIBLIOGRAPHY

BARA, S. et al., (1985), *Efectos ecológicos del Eucalyptus globulus en Galicia. Estudio comparative con Pinus pinaster y Quercus robur.* National Institute for Agricultural Research, Madrid.

CALS, J., CAPELLA, J., VAQUÉ, E., (1993), "El turismo en el desarrollo rural". Quasar, S.A. (Report prepared by the Ministry of Agriculture, Fisheries and Food. Madrid. Unpublished).

FAO, (1967), "World Symposium on Man-made Forests and their industrial importance". *Unasylva* 21 (3-4).

GANDULLO, J.M., SERRADA, R., (1993), "Problemática de los suelos agricolas de cara a su forestación. Tecnicas de preparacion del suelo". *Montes* 33.

GOMEZ MENDOZA, J., (1993), "Plantaciones forestales y restauración arbórea en Espana". *Revista de Occidente,* 149.

MADRIGAL, A., (1992), "Selvicultura Mediterránea: Una primera aproximación al tema". In I Jornada de Sevicultura Mediterranea. Grupo Ecologista Montes/E.T.S. Ingenieros des Montes Madrid.

MADRIGAL, A., (1994), *Ordenación de Montes Arbolados.* Serie Tecnica. National Institute for the Conservation of the Natural Environment. Madrid (in preparation).

OECD, (1991), *The State of the Environment.* Paris.

QUEZEL, P., (1977), "Los bosques de la cuenca mediterránea". In *Bosque y Maquia Mediterránea* (Tomaselli, R., Quezel, P. y Morandini, R.). Spanish translation by M. Crespo. Serval. Barcelona.

RUIZ DE LA TORRE, J., (1991), *Mapa Forestal de Espana Memoria General.* National Institute for the Conservation of the Natural Environment. Madrid.

SCHÜTZ, J.P., (1990) *Sylviculture 1: Principes d'éducation des forêts.* Presses Polytechniques et Universitaires Romandes. Lausanne.

SUMPSI, J.M., (1993) "Medio Ambiente y desarrollo forestal en el futuro del mundo rural". Quasar, S.A. (report prepared by the Ministry of Agriculture, Fisheries and Food. Madrid. Unpublished).

WIBE, S., (1992) *Market and Government Failures in Environmental Management. Wetlands and Forests.* OECD. Paris.

OFFICIAL STATEMENT

His Excellency Luis Maria Atienza Serna
Spanish Minister of
Agriculture, Fisheries and Food
Madrid

OFFICIAL STATEMENT

His Excellency Luis Maria Atienza Serna
Spanish Minister of
Agriculture, Fisheries and Food
Madrid

To begin with, I should like to thank all of you for coming here to the Madrid Conference Centre; my thanks also to the OECD and particularly to the Committee for Agriculture and the Environment Policy Committee, whose idea it was to hold this workshop. We shall be dealing with three subjects that must be discussed if we are to strike the necessary balance and achieve sustainable use of the planet's vital resources. Granted that agriculture can play its part in sustainable development only if it deals respectfully with the environment, forestry has to be considered an indispensable third element in achieving this balance. Such was the opinion expressed by the Spanish Delegation during the February 1994 session of the OECD Committee for Agriculture, so that we were very honoured that our application to host the workshop beginning here today was accepted.

The guidelines set out by the Earth Summit held in Rio de Janeiro in 1992 must be adhered to by all countries, although it seems unlikely that their implementation can be uniform. Since each country's situation is different, the guidelines will be applied differently, and there is a difficulty in trying to implement the same policies within the same time-frames. In my opinion, Rio should be seen as pointing the way forward or mapping out the line to be followed, but we must not stumble in our rush to achieve that goal.

As had already become evident in the European Community by the late 1980s, Europe's countryside is undergoing profound transformations. Many of them are caused by changes in external factors such as the balance of international trade in agricultural products, or by internal factors such as repeated surpluses owing to the unstoppable progress of agricultural technology, which is moving further and further away from traditional practices. At the same time, society is increasingly conscious of the degradation of the environment. More and more, city-dwellers are driven to seek relief from urban tensions in the countryside.

As a result, rural areas are undergoing profound changes, which have begun to transform their centuries-old patterns of life in dramatic ways. Because Spain is a Mediterranean country, its natural environment is extremely fragile, and country-dwellers as much as the environment around them are suffering ill effects.

Many of Spain's farms are still not big enough to be minimally viable for modern agriculture; in a great many cases their owners are more than 55 years old. In areas near large cities or thriving industrial complexes, many of these people must eke out their modest farm income with a second job completely unrelated to agriculture; in the most depressed areas, their only alternative, particularly for the youngest, is to uproot themselves and emigrate, as local prospects are too poor to encourage them to stay. Increasing numbers of landowners are faced with the dilemma of how to use their land. This is true not only of marginal land that will not yield much either when intensively farmed or used for grazing, but also of much other land, traditionally cropped or grazed, which is now being marginalised as well by adjustments under the Common Agricultural Policy.

If these lands are simply abandoned, and no corrective measures taken, there could be a worsening of the common plight of Mediterranean countries, which is particularly acute in Spain: erosion and the ensuing desertification, in addition to other environmental degradation. Forests are now recognised as having a threefold value -- economic, environmental and social -- and may represent a very attractive alternative to the abandonment of agriculture.

The majority of our forests are slow-growing and, although they do provide economic quantities of timber and fuelwood, they should also be seen as playing the valuable ecological role of stabilizing soils and regulating the natural water cycle. They are clearly more of an ecological than an economic asset, although I have the impression that this distinction is becoming increasingly difficult to make. The fact remains that the Mediterranean forest, which is unlike the woods of central and northern Europe in its appearance and mix of species, also serves other productive purposes, some of them of high economic and social value, such as the fruit and fodder production of our rangeland or hunting, which is the main economic use of woodlands in certain regions.

Spain's 50 million hectares can very roughly be said to be half woodland, half farmland. Because of Spain's long tradition of forestry, it has been possible to preserve down to the present day more than 6 million hectares of public and private woodland that has kept its productive and protective capacity and has remained fundamental to the economy of many rural communities.

It will be evident that to solve so complex a problem we shall need a strategy of sufficient scope to gather all of its threads together. The goal of the strategy should be to foster conservation of the natural environment through balanced and dynamic development of Spain's rural areas, and a fundamental thrust will be to reconcile the ecological, social and economic functions of our woodlands. The Ministry of Agriculture, Fisheries and Food has been guided by these principles in designing the National Strategy for the Overall Preservation of Nature, known by its Spanish acronym ENCINA, meaning "holm-oak". We hope that ENCINA, together with the Ministry's other rural revitalisation measures, will show how a model of sustainable development can best be achieved.

The Strategy has the three basic objectives of slowing environmental degradation through natural processes, maintaining biodiversity through conservation and the sustainable use of individual species, and reversing the effects of environmental degradation caused by human activity. Various approaches are being taken to achieve all three objectives and protect the land and sea environment.

Our intention in planting this "ENCINA" was that, like the rangeland tree of the same name, it would be the sturdy trunk from which the guidelines defining the Ministry's actions would branch off. We were not interested in making a conservation policy if it was to be just window dressing, showy but shallow; rather, our intention was that environmental concern should run through all our actions. Wonderful, well-preserved national parks are of no use if we allow aquifer-depleting activities to go on next door. Neither is there any point in maintaining marine reserves if nothing is done about overfishing. We can hardly say that we are protecting the environment if we cannot solve the agrifood industry's problems with waste products.

As part of the process now under way, our first draft of ENCINA will be discussed with the appropriate social actors -- essentially Autonomous Community governments, professional agrarian organisations, environmental movements and independent experts -- who will no doubt have valuable ideas and suggestions to make. Their contributions will help the Ministry to develop ENCINA, the sturdy tree with which we want to shelter all our actions.

In addition to these actions to maintain biodiversity through protection of the genetic resources of plant and animal species, I would like to say a few words about programmes now under way to combat desertification. Under a five-year investment plan, 450 000 hectares of eroded land is to be reforested, the plant cover restored on another 400 000 hectares, and forest fire prevention measures taken on 250 000 hectares. This five-year programme has a budget of Ptas 220 billion (US$1.8 billion), about Ptas 33 billion (US$0.3 billion) of which is to be spent in the first year.

This complements a programme begun last year as one of the so-called accompanying measures of the Common Agricultural Policy, consisting of income support for farmers who choose to plant marginal farmland to trees. Fifteen thousand farmers have so far applied, the area covered being some 330 000 hectares. Over the five-year life of the programme, 800 000 hectares are expected to be afforested.

To back up these initiatives, programmes are being implemented to preserve the genetic diversity of the principal forest species, with attention being paid to their genetic improvement and to their selection by geographical origin, so as to guarantee that, in afforestation and in the propagation of timber-yielding species and those of ecological value, quality seeds and seedlings of the appropriate ecotype are used.

We are convinced that all of these measures will bring about a profound change not only in rural structures but in the attitudes of rural inhabitants themselves, who for many years have been involved in a chapter of accidents not of their own making. After so many years of uncertainty as to their future, they require decisive assistance in terms of information, outreach and training from those at all levels who are responsible for policy implementation.

In addition, however, there is a need for urban society, with its growing demands on woodlands as places of leisure and relaxation, to become more involved in preserving them. City-dwellers should no longer be content with quick photo tours of the countryside, never straying far from their cars, but should become travellers, walkers, who understand and love the forest. There is little point in planting thousands of hectares every year if we cannot ensure that woodlands are respected by those who enjoy them. It is a commonplace to say that Spain is a "tree-killing" country. I refuse to believe it, and I am ready to work at making society understand the importance of forests and the need for everyone, managers and users, to protect them.

Society has only recently become concerned with environmental degradation, but in the countryside, people have long been worried about their future. What we are engaged in here is no mere academic exercise. Time is short and rural problems are pressing. I urge each and every one of you to do your utmost -- and I am confident that you will -- to see to it that this workshop brings us closer to solutions that will satisfy the demands made of us by society.

OFFICIAL STATEMENT

Gérard Viatte
Director for Food, Agriculture and Fisheries
OECD, Paris

OFFICIAL STATEMENT

Gérard Viatte
Director for Food, Agriculture and Fisheries
OECD, Paris

WELCOME

I wish first of all to thank our Spanish hosts most warmly for the very active role they have played in preparing and arranging this Workshop, and for their leading role in funding it. I should also like to thank six other financial contributors: the United States, the Commission of the European Communities, Finland, Japan, Portugal and Denmark. This OECD Workshop on Forestry, Agriculture and the Environment could not have been held here in Madrid without the generosity of these seven donors.

I further wish to thank all the experts on the Consultative Steering Group whose intellectual support has been most valuable, notably in selecting the main themes for the workshop. The OECD Secretariat also thanks all countries which have prepared case studies on their respective countries situations for this meeting. Last, thanks go to all of you who are here today: your participation augurs well for the coming discussions, and I am sure the workshop will benefit from your broad range of experience, both national and international. The considerable interest raised by the Workshop's themes is a pointer, I hope, to the success of the proceedings on which we are about to embark.

AGREEMENT ON THE REFORM OF
AGRICULTURAL POLICY IN OECD COUNTRIES

The necessity of reform. It had become clear over recent years that it was no longer possible to pursue policies of farm commodity price support backed by tariffs or quotas that encouraged agricultural output to rise in excess of consumption. This was evidenced by the strong downwards pressures which subsidised exports and import barriers exerted on world prices for agricultural commodities, by the accumulation of enormous stockpiles of many agricultural commodities, by the increasing burden on budgets and more generally by the rising costs to the economy as a whole.

These were the outward manifestations which compelled most countries to introduce corrective measures, beginning with measures to control supply. Policies to this effect were initially seen by many governments as the most politically acceptable and most pragmatic short-term solution to the problem of surpluses, given strong resistance to price cuts in many countries, which would have needed to be very substantial in some cases if they were to achieve any significant cutback in supply at a time when productivity was rising. But the measures to regulate supply continued in nearly every case to be linked to systems of price support and import restrictions, and proved inadequate to restore balance to the market.

The approach to reform. At its 1987 Meeting at Ministerial level the OECD Council drew attention to the serious problems in agricultural markets, and acknowledged that the main reason lay in support and protection policies that had prevented an adequate transmission of market signals to farmers. Ministers agreed that concerted reform of agricultural policies should be implemented in a balanced manner. The long-term objective was to allow market signals to influence the orientation of agricultural production by means of a progressive and concerted reduction of agricultural support, and by all other appropriate means. The Communiqué issued after the 1987 Ministerial Meeting sets out a comprehensive and integrated set of actions and principles for reforming agricultural policy in the context of structural adjustment.

To improve market orientation, agricultural policies need to be reformed so as to reduce production-related support and assistance to exports, improve market access and ensure that domestic prices are more fully aligned on world prices. Changes of this kind call for accompanying measures both in domestic markets and at the border. By contrast, one approach is to shift towards direct income support involving payments that are less strongly linked to production. These are more transparent, can be targeted more effectively, and may create less economic distortion than market price support. To make continued headway in this direction, however, the total level of assistance needs to be reduced, and border measures amended at the same time.

In spite of the significant changes that have occurred in recent years in the broad approach taken by OECD Member countries, the bulk of agricultural production is still essentially governed by support policies rather than responding to price signals from international markets. With few exceptions, the types of trade policies have remained as they were, making use of export subsidies and restrictions on access to domestic markets. This state of affairs continues to place extra cost burdens on domestic economies and distorts trade, as well as causing ongoing strains in international trading relations.

The necessity of reform is increasingly apparent, though to date the practical results have been modest. The principles set out in 1987 were reaffirmed by the OECD Agriculture Committee at its Ministerial Meeting in 1992 and its High-Level Meeting in early 1994. 1994 also saw the signature of the GATT Final Act in Marrakech, eight years after the Uruguay Round was launched. The agreements relating to agriculture cover a number of reforms involving binding commitments on market access, export subsidies and domestic assistance that distorts trade. These new disciplines for agriculture will bolster, in the OECD countries, the reform process moving towards greater market orientation, which should ultimately reduce the significant transfers from taxpayers and consumers to this sector.

IMPACT OF THE REFORM ON STRUCTURAL ADJUSTMENT, THE ENVIRONMENT AND RURAL DEVELOPMENT

One of the main reasons for governments' intervening in the agricultural sector is to safeguard the incomes of farmers and their families. The income objective is usually defined in terms of distribution or equity, taking as its basis the income of other population groups or the average income. But, transfers generated by market price support are directly proportional to the volume of output, and hence to the size of farms. A reduction in the level of these transfers, under agricultural policy reform, will thus have significant repercussions for the financial situation of large farms, to the extent that they draw a large share of their total income from agricultural activity alone. In the case of small farms, the impact of reform on overall income may be more modest, inasmuch as these farmers and their spouses have generally diversified their sources of income while continuing to farm part-time.

It seems likely that, to cope with the reforms, many farmers will endeavour to absorb (at least in the short term) the adverse effects on their income and to remain within their rural community, looking more to sources of income off the farm. The outcome of this transition will depend on the capacity of rural areas to generate new employment opportunities. Many of the agricultural labour force in many countries have relatively low levels of education; at the same time, farm workers are perhaps more flexible than others with regard to the spread of work over time, from season to season, and they are often used to working on their own, both factors that may be advantages for the development of rural communities.

There is a further dimension that is important here. The growing interest in the relationship between agriculture and the environment raises a number of problems that are significant for reform. The main questions have to do with consistency and compatibility in shaping policies for agriculture and the environment. For instance, considerable efforts are now being made to devise and introduce agro-environmental measures and programmes, yet in many countries these policies are set in a context of substantial agricultural support which contributes to generating adverse effects for the environment. However, although to date they represent only a minor part of overall farm support, over the last few years there has been an increase in direct payments for the provision of environmental services in agriculture, for both the physical environment (ecological contributions) and the social environment (cultural values). Such payments could well develop further in the future. These measures may usefully accompany the reform process, provided they are effectively targeted and are not directly linked to foodstuff production.

Agricultural policy reform and rural development are becoming continually more interdependent. Today, the well-being of many farm households depends increasingly on the non-agricultural income and employment opportunities available in rural areas. An important aspect of the reform process is to provide farmers and their families with opportunities to find other ways of increasing household incomes, unrelated directly to support for farm production. Relevant measures may include those for education and training, the provision of rural services, the promotion of specialised or niche markets, or the development of farm tourism. Rural policy may help facilitate the reform process by extending the range of opportunities for agricultural households to diversify by undertaking remunerative activity in other sectors.

THE WORKSHOP AND THE ACTIVITIES
OF THE OECD AGRICULTURE DIRECTORATE

At their meeting in March 1992, the Agriculture Ministers of the OECD countries set out fundamental orientations for the Agriculture Committee's work over the medium term. There were three main threads:

-- monitoring national policies and trends in markets and trade;

-- government action in the framework of the structural adjustment of agriculture;

-- analysis of matters relating to the environment and their consequences for government action.

It is the job of OECD to analyze and advance arguments in favour of the reform, and to demonstrate the longer term advantages for the farm sector and for the economy as a whole.

At the Workshop on Sustainable Agriculture Technology and Practices, held in February 1992 as part of the Technology and Environment Programme of the OECD Environment Directorate, forestry was identified as a significant element in any discussion of agriculture and the environment, and it was suggested that more time be devoted to that topic. Following on from the conclusions of that meeting, it was agreed to hold a Workshop on the interface between forestry, agriculture and the environment, drawing on voluntary contributions, under the guidance of the Joint Working Party of the Agriculture and Environment Committees. This recently established Working Party has been kept regularly informed of the progress of preparatory work and it should provide a useful framework for discussion of the main findings of this workshop at its forthcoming meeting in December 1994.

The present Workshop marks the beginning of OECD reflection on the interface between forestry, agriculture and the environment. This activity is of interest to policy makers, through its pragmatic and forward-looking policy oriented approach, and because it introduces a new and topical element into OECD work in analysing agricultural policies, including the agro-environmental dimension.

WHERE THE WORKSHOP MAY LEAD

The main point of interest is to attempt to see how far small-scale forestry has a role to play in the process of reforming agricultural policy, in timber production and in the protection of the environment. These are three key themes which deserve thorough consideration. While it is already apparent that the afforestation of land released from agriculture and the creation of forestry jobs for farmers cannot in themselves overcome all the problems of structural adjustment in agriculture, or the problems of the forestry economy and environmental protection, they may help to do so, and it is in identifying the specific circumstances in which that may happen that the workshop could make a valuable contribution to the debate.

If the workshop concludes that it would be beneficial to develop small-scale forestry in the OECD area, we shall have to consider *whether* policy incentives are needed and, if so, *what form* they should take. There can be no question of advocating poorly targeted policies to encourage small-scale forestry. Experience with agricultural policy in fact shows clearly that support of any kind, however justified it may be, has to meet specific objectives, minimising economic distortion and operating in favour of market-oriented production.

At their 1992 meeting, the OECD Agricultural Ministers stressed the growing importance of the two-way relationships between agriculture and forestry and the environment. In a broad approach, it would be useful to clarify the nature of the public goods which forestry provides, and to take stock of current thinking about remunerating the non-market functions of forestry stands which are beneficial to the natural environment and to society. More specifically, the workshop could well consider the selection or range of policy instruments and approaches to be recommended to promote small-scale forestry of an environmentally-friendly kind, in the context of the reform of agricultural policy.

In relation to the forestry economy, the progressive release of land from farming in some parts of the OECD area may be seen as an opportunity to develop forestry production. In a broad approach, it would be useful to consider trends and outlook in markets for forestry products, and the possible developments in the timber balance in OECD countries. More specifically, the workshop could well take a preliminary view of how, and how far, afforestation of agricultural land may supplement, or even replace, in some cases, traditional forestry activity in terms of timber outlets, with all the possible implications for forestry and timber industry policy.

Current trends in agriculture show the growing importance of efforts to generate sources of employment and income for farmers outside their sector. Forestry may be a good means of helping to develop disadvantaged rural areas which are becoming depopulated but may well be suitable for timber production. It may help make farms more viable, via remuneration for work in other people's woodlands or the sale of wood and timber. The workshop may lay foundations for analysis of the role of small-scale forestry in the framework of the rural policies introduced by OECD countries.

To conclude, I wish the workshop all success. I trust that the discussions will be both lively and productive, and that the workshop will help to provide a number of useful signposts and pointers for the discussion and development of policy relating to forestry, agriculture and the environment.

OFFICIAL STATEMENT

Georges Touzet
Director General of the French National Forest Agency
Paris

OFFICIAL STATEMENT

Georges Touzet
Director General of the French National Forest Agency
Paris

The previous papers have described how and why this workshop was conceived by the OECD. If I may summarise what has been said, agricultural policy reform in OECD countries will progressively lead to changes in the use of farmland, in farm incomes and in rural employment. Forestry has been put forward by governments as an option in the context of reform, both to produce wood, thus generating income, and to respond to increasing concerns about the quality of the environment.

The issue has been outlined for us and the expectations are clear. In response, members of the workshop should first of all ensure that their proposals are feasible from a practical point of view, bearing in mind that the problem of farm forestry is not entirely new -- in Europe at least -- and that there are countries with some experience of it. They should also remember that the timber market and agricultural commodity markets are two different things. Finally, they should not overlook the constraints -- and their costs -- that stem from environment policy.

I should like to say a few words on each of these three last points.

Past experience

"Agricultural abandonment" is a trend that has long existed in Europe; it began with the two World Wars and has gathered pace over the past 40 years. Some of the land concerned has been naturally or artificially turned over to woodland, and some left fallow.

In France, reforestation was one of the objectives set by the National Forestry Fund, thanks to which some two to three million hectares have been planted with trees.

More recently, at any rate up until the reform of the Common Agricultural Policy (CAP), the European Economic Community included reforestation in its structural adjustment policy for agriculture. This kind of small-scale forestry work was granted financial aid from the European Agricultural Guidance and Guarantee Fund (EAGGF).

There are other examples, but these two alone suggest that the issue has been a concern for some time and that sufficient practical steps have been taken for a few lessons to be learned.

In fact two problems arose. First, areas afforested were small and scattered, and second, it took a long time before farmers derived any income from the woodland, not to mention that such income was also very low compared with farm earnings.

The division of farms into small plots led to "postage-stamp" reforestation, which encroached on the remaining farmland and produced stands of trees too small to be managed on a sustainable basis. This kind of reforestation may even have encouraged more farmers to leave the land and hastened the drift from the countryside. In the absence of planning, it may also have blighted the landscape.

It is all the harder for farmers to accept the long payback period before deriving any income from woodland in that their initial investment is far from insignificant.

In France, for instance, attempts were made by smallholders to join forces and set up *Groupements forestiers* (joint forestry ventures). The outcome was inconclusive. Ultimately, reforestation was either on a very small scale or on larger areas of land purchased by financial or industrial investors, but few of them being farmers.

Without substantial investment grants, no reforestation would have occurred at all. In some cases, it was combined with an annual subsidy paid over several years to offset the loss of income to farmers. This was not enough to generate a significant trend towards farm forestry.

To reduce the payback period, short-rotation stands were proposed, in particular short-rotation coppices. This is a technique commonly used in some countries to supply pulp factories. The problem then arises of how to guarantee that the output will be fully cleared on the market.

This is a particularly serious issue in short-rotation forestry, as it is in general. Of course wood is and should remain a raw material that is inexpensive, and such a product thus cannot incur excessively high transport costs. Any policy to plant new stands for production must or should ideally be supported by a policy to create or expand user industries.

Timber market

The timber market differs from agricultural commodity markets. In particular, it is highly sensitive to the general economic climate and, as a free global market, to fluctuations in the currency markets. At times, such as the period we have just been through, competition becomes extremely keen, all the more so because some countries are naturally well-endowed with woodland, dense forest cover, highly productive woodland and low forestry/harvesting costs. It is obvious, for instance, that eucalyptus production in Iberia bears little relation to that in Brazil, and similar examples abound.

Proposals have been put forward regarding an international division of timber production, but this would be of course of very little benefit to farm forestry.

While no panacea is at hand, there is a clear need to look into forms of production that may have a comparative advantage in trade. There is also a need for forestry techniques that allow harvesting to be brought forward or postponed for two or three years without any loss to the farmer.

Environment

Afforestation can blight the landscape when views are blocked or plots of woodland are too small. Some softwood plantations, especially when thinning is inadequate, acidify both soil and water. Reforestation policy must therefore involve rules and constraints to prevent such damage.

Moreover, requirements relating to landscape, biodiversity and cutting push up the cost of forestry and harvesting. In some cases they may reduce the production capacity of woodland. Frequent mention is made about the requirements imposed on foresters, but seldom about payment for the services they render.

Of course there is the cost of public access to very popular areas of woodland that produce virtually no timber at all, but this is never likely to apply to farm forestry.

However, some farmland and pastureland, once planted with trees, undeniably play a protective role against erosion and avalanches, with timber production always being a secondary activity, if only on grounds of cost. This is a service to the community which should be remunerated at least when the woodland is under private ownership.

Mention has been made of specialised forests, some for conservation, some for recreational use, and others for timber production. From a general viewpoint this is questionable, but with respect to farm forestry it is quite simply unacceptable until and unless a way is found of setting a value on services to the environment, conservation and public access and establishing a reliable system of remuneration.

Conclusion

The OECD initiative is timely, coming neither too early -- before countries have been able to acquire some experience of the technical, social and economic issues involved -- nor too late.

The OECD is providing an opportunity to express opinions and give advice. Those who are involved at grassroots level, with a practical grasp of the technical, financial and commercial problems involved must seize this chance. While the discussions are practical and rather down-to-earth, the advice that will be passed on to policy-makers will be relevant to their concerns. They in turn will have to avoid the temptation of becoming too bureaucratic and bear in mind that, in this field more than in many others, there is nothing worse than seeing things as we think they ought to be seen!

OFFICIAL STATEMENT

Mr. Carlos Tió Saralegui,
Secretary General for Agrarian Structures,
Spanish Ministry of Agriculture, Fisheries and Food
Madrid

OFFICIAL STATEMENT

Mr. Carlos Tió Saralegui,
Secretary General for Agrarian Structures,
Spanish Ministry of Agriculture, Fisheries and Food
Madrid

I should like to remind our friends from so many countries who have attended this workshop of a few aspects of Spain's history which illustrate the troubled relations between agriculture, stockbreeding and forestry.

For centuries a stockbreeders' guild, El Honrado Concejo de la Mesta, held sway over arable farmers, substantially inhibiting the development of agriculture. There was a ban on the enclosure of agricultural land, and the movement of herds and flocks devastated crops.

Farmers too, with large-scale clearances of woodland in periods of food scarcity and the struggle for survival, made suicidal attempts to impose their will on nature, again for centuries.

Lastly, it should be noted how, much more recently, in the present century, the reafforestation programmes decreed by government drove the stockbreeders off their finest grazing areas, stimulating farmers' hatred for woodlands.

These are just three examples of land-use competition between agriculture, stockbreeding and forestry which have marked the history of Spain and no doubt many other countries too.

The position now is of course very different, and we can look to fresh and harmonious relations between these rural activities, though we would do well not to forget the past.

The two most striking factors that have changed the earlier scenario of confrontation are:

a) the surplus of food supplies in many industrialised countries, leading to saturation on world markets for these commodities;

b) the growing concern for the environment, which has spread through most advanced societies.

The result has been to disturb the conception of the agrarian and rural world that has so far prevailed over the second half of the 20th century in the most industrialised countries, which we may identify as covering the membership of OECD.

The idea of a new model for rural development based on a broader combination of complementary activities has made headway in recent years. Under this model, agriculture will continue to be the basic economic activity in the majority of rural areas. But agriculture will no longer aim solely at unlimited increases in yields. Markets are beginning to attach value to further quality-related aspects, while control of production costs may in future become a basic factor in farm profitability against the new background of farm prices steadily aligning themselves with those on world markets.

While agriculture is a basic activity in the rural economy it is not, and cannot be, sufficient to sustain the whole of that economy. Agriculture has reached its limits in generating income for the rural population, and the use of resources -- land, labour and capital -- must be diversified within a new and more balanced model of integrated rural development.

All this is against the backdrop, in the industrialised countries, of fresh and unsatisfied demands for leisure in the countryside, a need to move away from large conurbations and recover the natural environment that has gradually been deteriorating to the point where it might well disappear forever.

The population of the rural world can thus once again provide a service of enormous value for society as a whole. Here it will be a matter of restoring the balanced use of resources as between agriculture, stockbreeding, forestry, environmental conservation and biodiversity.

This is no easy task; nor should it be regarded as a romantic return to the past. In my opinion, the new model of rural development is an historical necessity and needs to be advanced with due regard for market forces and the principles of competition that guide our economies. The new activities to be built up in the rural world must respond to genuine demand among the population; they must have costs and prices, and generate stable income. Otherwise we should be bringing forth a new lame-dog sector, permanently dependent on government assistance.

Government assistance is no doubt necessary in this transitional phase, where new markets (for rural tourism, for example) have to be developed or because, in the switch from an economy that spoils the environment to one based on sustainable development, the market still continues to allocate prices and costs without attaching proper weight to the environmental effects of production processes. During the transition, government intervention via subsidies, bans and penalties is decisive.

There are many other aspects which could usefully be emphasized, and I would mention in the first place that the change that we are witnessing is necessarily gradual and has to be international.

The slow and gradual nature of this change to a new model must be understood not as any wish to inhibit the process, which is clearly irreversible; the reason for it lies in the need for far-reaching adjustment, including education systems, given that the new awareness of nature and the sustainable development of rural areas calls for a generation change, both on the part of users and on the part of the population of the rural world, in many cases now ageing, but who earlier experienced the marvels of the green revolution and came to terms with the laws of productivity.

Nor should it be forgotten that, if it is to prove successful, this change of models for the rural economy must be international; otherwise protectionist barriers will again be erected. The growing openness of world markets stands no chance of proving stable and permanent unless shared practices and standards on environmental, social, health, fiscal and monetary matters are widely adopted.

Otherwise we should be triggering off the wildest form of capitalism in the world's history, and signing the death warrant of the social and economic achievements of western civilisation.

WORKSHOP ON FORESTRY, AGRICULTURE AND THE ENVIRONMENT

MADRID, 17-20 OCTOBER, 1994

LIST OF PARTICIPANTS

ALLEMAGNE Mr. Uwe VANSELOW
GERMANY Ambassade d'Allemagne
Madrid

AUTRICHE/ Mr. Helmut WALTER
AUSTRIA Federal Ministry of Agriculture
and Forestry

Mrs. Karla KRIEGER
Federal Ministry of Agriculture
and Forestry

BELGIQUE/ M. Marcel LAFARGE
BELGIUM Ambassade de Belgique
Madrid

DANEMARK/ Mrs. Ditte SVENDSEN
DENMARK Danish Forest and Landscape
Research Institute

Ms. Gertrud KNUDSEN
Ministry of the Environment

Mr. Bjarne THOMSEN
Ministry of Agriculture and Fisheries

ESPAGNE/ M. Luis Maria ATIENZA SERNA
SPAIN Ministro de Agricultura, Pesca
 y Alimentacion

M. Carlos TIO SARALEGUI
Ministerio de Agricultura, Pesca
y Alimentacion

M. Joaquin CASTILLO SEMPERE
Ministerio de Agricultura, Pesca
y Alimentacion

M. Fernando ESTIRADO GOMEZ
Instituto Nacional para la
Conservacion de la Naturaleza
(ICONA)

M. Fernando GOMEZ JOVER
Ministerio de Agricultura, Pesca
y Alimentacion

M. Angel BARBERO
Instituto Nacional para la
Conservacion de la Naturaleza
(ICONA)

M. Rafael ALVAREZ RODRIGUEZ
Ministerio de Agricultura, Pesca
y Alimentacion

M. Francisco J. JIMENEZ PERIS
Ministerio de Agricultura, Pesca
y Alimentacion

M. Jose Luis MILAS CLIMENT
Instituto Nacional de Investigacion
y Tecnologica Agraria y Alimentaria
(INIA)

M. Antonio PARAMIO DURAN
Ministerio de Obras Publicas,
Transporte y Medio Ambiente

M. Jesus GONZALEZ REGIDOR
Délégation de l'Espagne
près l'OCDE

M. Arturo ACOSTA GARRIDO
Estructuras y Extension Agraria
Consejeria de Economia
Madrid

M. Jesus ALVAREZ ARAGONESES
Consejeria de Medio Ambiente
Junta de Castilla y Leon

M. Miguel TRONCOSO HERMOSO
DE MENDOZA
Aplicacion y Seguimiento PAC
Gobierno de Navarra

M. Felix OCHOA BRIZUELA
Consejeria Medio Rural y Pesca
Principado de Asturias

M. Jose ESCORIHUELA MESTRE
Direccion General del Medio Natural
Barcelona

ETATS-UNIS/ Mr. John A. MIRANOWSKI
UNITED STATES   U.S. Department of Agriculture

Mr. Peter F. SMITH
U.S. Department of Agriculture

FINLANDE/ Mr. Ilkka VAINIO-MATTILA
FINLAND Ministry of Agriculture and Forestry

Mr. Niilo HINTIKKA
Ministry of Agriculture and Forestry

Mr. Jan HEINO
Ministry of Agriculture and Forestry

Mrs. Leena KARJALAINEN-BALK
Ministry of the Environment

Mr. Ilpo TIKKANEN
University of Helsinki
Department of Forest Economics

Mr. Birger SOLBERG
European Forest Institute

Mr. Pentti HYTTINEN
European Forest Institute

FRANCE   M. Georges TOUZET
Office National des Forêts

M. Guy POIRIER
Ministère de l'agriculture et de la pêche

GRECE/   Mr. Yannis PETAMIDES
GREECE   Ministry of Agriculture

IRLANDE/   Mr. Tom POWER
IRELAND   Department of Agriculture,
   Food and Forestry

Ms. Eilish KENNEDY
Department of Agriculture,
   Food and Forestry

ITALIE/   M. Silvano SALVATICI
ITALY   Ministère des Ressources Agricoles,
Alimentaires et Forestières

M. Paolo VICENTINI
Ministère des Ressources Agricoles,
Alimentaires et Forestières

JAPON/   Mr. Takeshi GOTO
JAPAN   Ministry of Agriculture, Forestry and Fisheries

Mr. Masahiko HORI
Ministry of Agriculture, Forestry and Fisheries

Mr. Makoto OSAWA
Delegation of Japan to the OECD

MEXIQUE/ **MEXICO**	Ms. Beatriz AVALOS SARTORIO Ambassade du Mexique Bruxelles
NOUVELLE- **ZÉLANDE/** **NEW ZEALAND**	Mr. John VALENTINE Chief Executive Secretary Ministry of forestry
	Mr. John JACKMAN New Zealand Pastoral Agriculture Research Institute Ltd.
	Mr. Ken SHIRLEY New Zealand Forest Owners' Association Inc.
	Mr. M.E.F. SMITH New Zealand Farm Forestry Association
PAYS-BAS/ **NETHERLANDS**	Mr. R.A. ZAKEE Staff Officer Forestry Policy Natuur, Bos, Landschap en Fauna
	Mr. C.F.W.M. von MEIJENFELDT Staff Officer Forestry Policy Natuur, Bos, Landschap en Fauna
PORTUGAL	M. Carlos José MORAIS Instituto Forestal
	Mme Conceiçao FERREIRA Instituto Forestal
	M. Antonio Alberto GONÇALVES FERREIRA Conselho Nacional das Florestas/CAP
	M. Armando FIALHO CELPA

M. Pedro FERREIRA
Association des Industries et
Exportateurs de Cortiça do Norte

ROYAUME-UNI/   Mr. Neil CUMBERLIDGE
UNITED KINGDOM   Ministry of Agriculture, Fisheries
and Food

Mr. Willie SHERIDAN
Forestry Commission

Mr. Howard FEARN
Ministry of Agriculture, Fisheries and Food

Ms. Sheila McCABE
Department of Environment

SUEDE/   Ms. Maria LARSSON
SWEDEN   Ministry of Agriculture

Mr. Svante LUNDQUIST
Ministry of Environment
and Natural Resources

Mr. Carl Johan LIDEN
Swedish Board of Agriculture

Mr. Gunnar NORDANSTIG
National Board of Forestry

SUISSE/   M. Daniel ZÜRCHER
SWITZERLAND   Office Fédéral de l'Environnement,
des Forêts et du Paysage

M. David SCHMID
Office Fédéral de l'Environnement,
des Forêts et du Paysage

COMMUNAUTÉS EUROPÉENNES/
EUROPEAN COMMUNITIES

M. Robert FLIES
Unité actions spécifiques
en milieu rural
Commission des Communautés Européennes

PAYS AYANT UN STATUT D'OBSERVATEUR AUPRÈS DU COMITÉ DE L'AGRICULTURE DE L'OCDE/OBSERVER COUNTRIES TO THE OECD COMMITTEE FOR AGRICULTURE

HONGRIE/
HUNGARY

Mr. Ferenc NYUJTO
Ministry of Agriculture

Mr. Gyula HOLDAMPF
Ministry of Agriculture

POLOGNE/
POLAND

Ms. Izabela KAKOL
Ministry of Environmental Protection,
Natural Resources and Forestry

REPUBLIQUE SLOVAQUE/
SLOVAK REPUBLIC

Mr. J. KONOPKA
Ministry of Agriculture

Mrs. P. GALOVA
Ministry of Agriculture

Mr. J. NOVOTNY
Forest Research Institute

REPRÉSENTANTS DES NATIONS UNIES ET INSTITUTIONS SPÉCIALISÉES/ REPRESENTATIVES OF THE UNITED NATIONS AND SPECIALISED AGENCIES

COMMISSION ÉCONOMIQUE POUR L'EUROPE DES NATIONS UNIES/ORGANISATION DES NATIONS UNIES POUR L'ALIMENTATION ET L'AGRICULTURE/ UNITED NATIONS ECONOMIC COMMISSION FOR EUROPE/ FOOD AND AGRICULTURE ORGANISATION OF THE UNITED NATIONS

Mr. C.F.L. PRINS
Agriculture and Timber Division
ECE/FAO

ORGANISATIONS INTERGOUVERNEMENTALES/ INTERGOVERNMENTAL ORGANISATIONS

CENTRE INTERNATIONAL DE HAUTES ETUDES AGRONOMIQUES MÉDITERRANÉENNES / INTERNATIONAL CENTER FOR ADVANCED MEDITERRANEAN AGRONOMIC STUDIES

M. Miguel VALLS ORTIZ
Institut Agronomique Méditerranéen,
Saragosse

M. Guillermo FLICHMAN
Institut Agronomique Méditerranéen,,
Montpellier

M. Placido PLAZA
CIHEAM,
Paris

ORGANISATIONS INTERNATIONALES NON GOUVERNEMENTALES/ INTERNATIONAL NON-GOVERNMENTAL ORGANISATIONS

CONFÉDÉRATION EUROPÉENE POUR L'AGRICULTURE/ EUROPEAN CONFEDERATION OF AGRICULTURE

Mr. Johannes SCHIMA
Austrian Committee for
Agriculture and Forestry

M. Miguel PEREZ TURADO
Union des Sylviculteurs du Sud
de l'Europe

FÉDÉRATION INTERNATIONALE DES PRODUCTEURS AGRICOLES/ INTERNATIONAL FEDERATION OF AGRICULTURAL PRODUCERS

Ms. Frances KINNON
Secretary for Rural Development

CONSULTANTS

Mr. J.R. CRABTREE
Macaulay Land Use Research Institute
Aberdeen

M. Alberto MADRIGAL COLLAZO
Escuela Superior Tecnica de
Ingenieros de Montes
Madrid

Mr. Peter H. PEARSE
Faculty of Forestry,
The University of British Columbia
Vancouver

Mr. Tim J. PECK
European Forest Institute
Joensuu

SECRÉTARIAT DE L'OCDE/OECD SECRETARIAT

DIRECTION DE L'ALIMENTATION, DE L'AGRICULTURE ET DES PÊCHERIES/
DIRECTORATE FOR FOOD, AGRICULTURE AND FISHERIES

M. Gérard VIATTE
Directeur

M. Wilfrid LEGG
Chef de la Division
Division Etudes Nationales II et
Environnement

M. Gérard BONNIS
Administrateur
Division Etudes Nationales II et
Environnement

Mme. Françoise BENICOURT
Secrétaire
Division Etudes Nationales II et
Environnement

Mme. Flavia GIROUARD
Secrétaire
Division Etudes Nationales II et
Environnement

DIRECTION DE L'ENVIRONNEMENT/
ENVIRONMENT DIRECTORATE

M. Thomas JONES
Administrateur
Division de l'Economie

MAIN SALES OUTLETS OF OECD PUBLICATIONS
PRINCIPAUX POINTS DE VENTE DES PUBLICATIONS DE L'OCDE

ARGENTINA – ARGENTINE
Carlos Hirsch S.R.L.
Galería Güemes, Florida 165, 4° Piso
1333 Buenos Aires   Tel. (1) 331.1787 y 331.2391
Telefax: (1) 331.1787

AUSTRALIA – AUSTRALIE
D.A. Information Services
648 Whitehorse Road, P.O.B 163
Mitcham, Victoria 3132   Tel. (03) 873.4411
Telefax: (03) 873.5679

AUSTRIA – AUTRICHE
Gerold & Co.
Graben 31
Wien I   Tel. (0222) 533.50.14
Telefax: (0222) 512.47.31.29

BELGIUM – BELGIQUE
Jean De Lannoy
Avenue du Roi 202 Koningslaan
B-1060 Bruxelles   Tel. (02) 538.51.69/538.08.41
Telefax: (02) 538.08.41

CANADA
Renouf Publishing Company Ltd.
1294 Algoma Road
Ottawa, ON K1B 3W8   Tel. (613) 741.4333
Telefax: (613) 741.5439
Stores:
61 Sparks Street
Ottawa, ON K1P 5R1   Tel. (613) 238.8985
211 Yonge Street
Toronto, ON M5B 1M4   Tel. (416) 363.3171
Telefax: (416)363.59.63

Les Éditions La Liberté Inc.
3020 Chemin Sainte-Foy
Sainte-Foy, PQ G1X 3V6   Tel. (418) 658.3763
Telefax: (418) 658.3763

Federal Publications Inc.
165 University Avenue, Suite 701
Toronto, ON M5H 3B8   Tel. (416) 860.1611
Telefax: (416) 860.1608

Les Publications Fédérales
1185 Université
Montréal, QC H3B 3A7   Tel. (514) 954.1633
Telefax: (514) 954.1635

CHINA – CHINE
China National Publications Import
Export Corporation (CNPIEC)
16 Gongti E. Road, Chaoyang District
P.O. Box 88 or 50
Beijing 100704 PR   Tel. (01) 506.6688
Telefax: (01) 506.3101

CHINESE TAIPEI – TAIPEI CHINOIS
Good Faith Worldwide Int'l. Co. Ltd.
9th Floor, No. 118, Sec. 2
Chung Hsiao E. Road
Taipei   Tel. (02) 391.7396/391.7397
Telefax: (02) 394.9176

**CZECH REPUBLIC – RÉPUBLIQUE
TCHÈQUE**
Artia Pegas Press Ltd.
Narodni Trida 25
POB 825
111 21 Praha 1   Tel. 26.65.68
Telefax: 26.20.81

DENMARK – DANEMARK
Munksgaard Book and Subscription Service
35, Nørre Søgade, P.O. Box 2148
DK-1016 København K   Tel. (33) 12.85.70
Telefax: (33) 12.93.87

EGYPT – ÉGYPTE
Middle East Observer
41 Sherif Street
Cairo   Tel. 392.6919
Telefax: 360-6804

FINLAND – FINLANDE
Akateeminen Kirjakauppa
Keskuskatu 1, P.O. Box 128
00100 Helsinki
Subscription Services/Agence d'abonnements :
P.O. Box 23
00371 Helsinki   Tel. (358 0) 121 4416
Telefax: (358 0) 121.4450

FRANCE
OECD/OCDE
Mail Orders/Commandes par correspondance:
2, rue André-Pascal
75775 Paris Cedex 16   Tel. (33-1) 45.24.82.00
Telefax: (33-1) 49.10.42.76
Telex: 640048 OCDE
Internet: Compte.PUBSINQ @ oecd.org
Orders via Minitel, France only/
Commandes par Minitel, France exclusivement :
36 15 OCDE

OECD Bookshop/Librairie de l'OCDE :
33, rue Octave-Feuillet
75016 Paris   Tel. (33-1) 45.24.81.81
(33-1) 45.24.81.67

Documentation Française
29, quai Voltaire
75007 Paris   Tel. 40.15.70.00

Gibert Jeune (Droit-Économie)
6, place Saint-Michel
75006 Paris   Tel. 43.25.91.19

Librairie du Commerce International
10, avenue d'Iéna
75016 Paris   Tel. 40.73.34.60

Librairie Dunod
Université Paris-Dauphine
Place du Maréchal de Lattre de Tassigny
75016 Paris   Tel. (1) 44.05.40.13

Librairie Lavoisier
11, rue Lavoisier
75008 Paris   Tel. 42.65.39.95

Librairie L.G.D.J. - Montchrestien
20, rue Soufflot
75005 Paris   Tel. 46.33.89.85

Librairie des Sciences Politiques
30, rue Saint-Guillaume
75007 Paris   Tel. 45.48.36.02

P.U.F.
49, boulevard Saint-Michel
75005 Paris   Tel. 43.25.83.40

Librairie de l'Université
12a, rue Nazareth
13100 Aix-en-Provence   Tel. (16) 42.26.18.08

Documentation Française
165, rue Garibaldi
69003 Lyon   Tel. (16) 78.63.32.23

Librairie Decitre
29, place Bellecour
69002 Lyon   Tel. (16) 72.40.54.54

Librairie Sauramps
Le Triangle
34967 Montpellier Cedex 2   Tel. (16) 67.58.85.15
Tekefax: (16) 67.58.27.36

GERMANY – ALLEMAGNE
OECD Publications and Information Centre
August-Bebel-Allee 6
D-53175 Bonn   Tel. (0228) 959.120
Telefax: (0228) 959.12.17

GREECE – GRÈCE
Librairie Kauffmann
Mavrokordatou 9
106 78 Athens   Tel. (01) 32.55.321
Telefax: (01) 32.30.320

HONG-KONG
Swindon Book Co. Ltd.
Astoria Bldg. 3F
34 Ashley Road, Tsimshatsui
Kowloon, Hong Kong   Tel. 2376.2062
Telefax: 2376.0685

HUNGARY – HONGRIE
Euro Info Service
Margitsziget, Európa Ház
1138 Budapest   Tel. (1) 111.62.16
Telefax: (1) 111.60.61

ICELAND – ISLANDE
Mál Mog Menning
Laugavegi 18, Pósthólf 392
121 Reykjavik   Tel. (1) 552.4240
Telefax: (1) 562.3523

INDIA – INDE
Oxford Book and Stationery Co.
Scindia House
New Delhi 110001   Tel. (11) 331.5896/5308
Telefax: (11) 332.5993
17 Park Street
Calcutta 700016   Tel. 240832

INDONESIA – INDONÉSIE
Pdii-Lipi
P.O. Box 4298
Jakarta 12042   Tel. (21) 573.34.67
Telefax: (21) 573.34.67

IRELAND – IRLANDE
Government Supplies Agency
Publications Section
4/5 Harcourt Road
Dublin 2   Tel. 661.31.11
Telefax: 475.27.60

ISRAEL
Praedicta
5 Shatner Street
P.O. Box 34030
Jerusalem 91430   Tel. (2) 52.84.90/1/2
Telefax: (2) 52.84.93

R.O.Y. International
P.O. Box 13056
Tel Aviv 61130   Tel. (3) 546 1423
Telefax: (3) 546 1442

Palestinian Authority/Middle East:
INDEX Information Services
P.O.B. 19502
Jerusalem   Tel. (2) 27.12.19
Telefax: (2) 27.16.34

ITALY – ITALIE
Libreria Commissionaria Sansoni
Via Duca di Calabria 1/1
50125 Firenze   Tel. (055) 64.54.15
Telefax: (055) 64.12.57
Via Bartolini 29
20155 Milano   Tel. (02) 36.50.83

Editrice e Libreria Herder
Piazza Montecitorio 120
00186 Roma   Tel. 679.46.28
Telefax: 678.47.51

Libreria Hoepli
Via Hoepli 5
20121 Milano   Tel. (02) 86.54.46
Telefax: (02) 805.28.86

Libreria Scientifica
Dott. Lucio de Biasio 'Aeiou'
Via Coronelli, 6
20146 Milano   Tel. (02) 48.95.45.52
Telefax: (02) 48.95.45.48

JAPAN – JAPON
OECD Publications and Information Centre
Landic Akasaka Building
2-3-4 Akasaka, Minato-ku
Tokyo 107   Tel. (81.3) 3586.2016
Telefax: (81.3) 3584.7929

KOREA – CORÉE
Kyobo Book Centre Co. Ltd.
P.O. Box 1658, Kwang Hwa Moon
Seoul   Tel. 730.78.91
Telefax: 735.00.30

OECD PUBLICATIONS, 2 rue André-Pascal, 75775 PARIS CEDEX 16
PRINTED IN FRANCE
(51 95 12 1) ISBN 92-64-14580-X - No. 48185 1995